Purebred Livestock Marketing

2nd Edition

A complete guide to livestock advertising and promotion.

Address editorial correspondence to:

Rachel Cutrer
1730 North Richmond
Wharton, Texas 77488
979-532-9141
979-532-9977 (fax)
rachel@ranchhousedesigns.com

Cover photo and author photo by Luke Neumayr, www.lukeandcat.com
Back cover photos by John Reasor (Argentina Angus Show), Paul Wayre (showpig) and Melissa Grimmell (bull).

Published March 2014 by Tops Printing, Bryan, TX
ISBN # 978-0-615-97652-5
Copyright ©2011, 2014 by Rachel Cutrer.

Printed in the United States of America.

TABLE OF CONTENTS

CHAPTER 1:

Marketing Purebred Livestock

We can love them, breed them, promote them and show them, but if we cannot market these cattle, we cannot survive. - Carlos X. Guerra

The above statement effectively explains the importance of marketing purebred livestock at the most basic level. As economics force change in the purebred livestock business, many producers are looking for ways to both cut expenses and increase their sales prices in order to survive.

Marketing is crucial to any business, but it seems to be especially important in the purebred livestock industries, because in many instances the eventual sales price of an animal is based upon not only the animal's quality , but also the marketing and promotional work done by the breeder.

The livestock business is unique, because it is a business, it is also an enjoyable activity, and thus sometimes viewed as a hobby for some, and a business for others. However, it is crucially important to understand that the livestock business is a business just like any other company, and a business must be profitable to be a success. While it is very enjoyable to own livestock, at the end of the day, the livestock should be able to generate a profit.

Before beginning any marketing program it is important to understand the difference between marketing and promotion. Marketing is a strategic process. It is carefully planned, implemented and evaluated. Marketing purebred livestock, as with any product, is a science based on many factors: consumer behavior, the economy, and much more. Marketing is not completed until you have a completed sales transaction with the check cleared in the bank.

Promotion is one of the four components of a marketing program. Marketing professionals will commonly refer to the four "P's" of marketing: product, price, place, and promotion. Promotion is fun. Promotion is exciting. Promotion engages your clients with your operation and draws attention to your brand. It requires constant, ongoing effort to try new things and improve on your current efforts.

There are more people involved in the sales, marketing, and promotion of purebred livestock today than ever before. As competition for sales gets tougher, purebred livestock producers must utilize more of the marketing and promotion functions in order to be successful.

Marketing Purebred vs. Commodity Livestock

Commercial livestock marketing usually occurs as a fairly standard procedure with the price based on general characteristics such as weight, age, color, and breed type. Marketing purebred livestock is completely different than marketing commercial livestock. In marketing purebred livestock, the phenotypic quality of the animal serves as a first impression to draw the buyer in, however, other traits such as pedigree, disposition, and more, all are characteristics of a quality sale animal.

In the purebred livestock sector, producers have the opportunity to greatly increase the price point of their livestock based on their own promotional efforts. As outlined in the beef industry funnel, marketing options for producers occur both on the farm and off the farm. For seedstock or purebred producers, and even cow/calf producers, the individual producer has the power to set their pricing and determine their market. This is where promotional efforts can aid in your marketing. Once the animal leaves the farm and moves to the feedlot, meat packing, retailer, or consumer level - as in what most commercial beef producers do -- there is no opportunity to increase your price significantly through your promotional efforts.

The purebred livestock marketing process takes on a unique identity

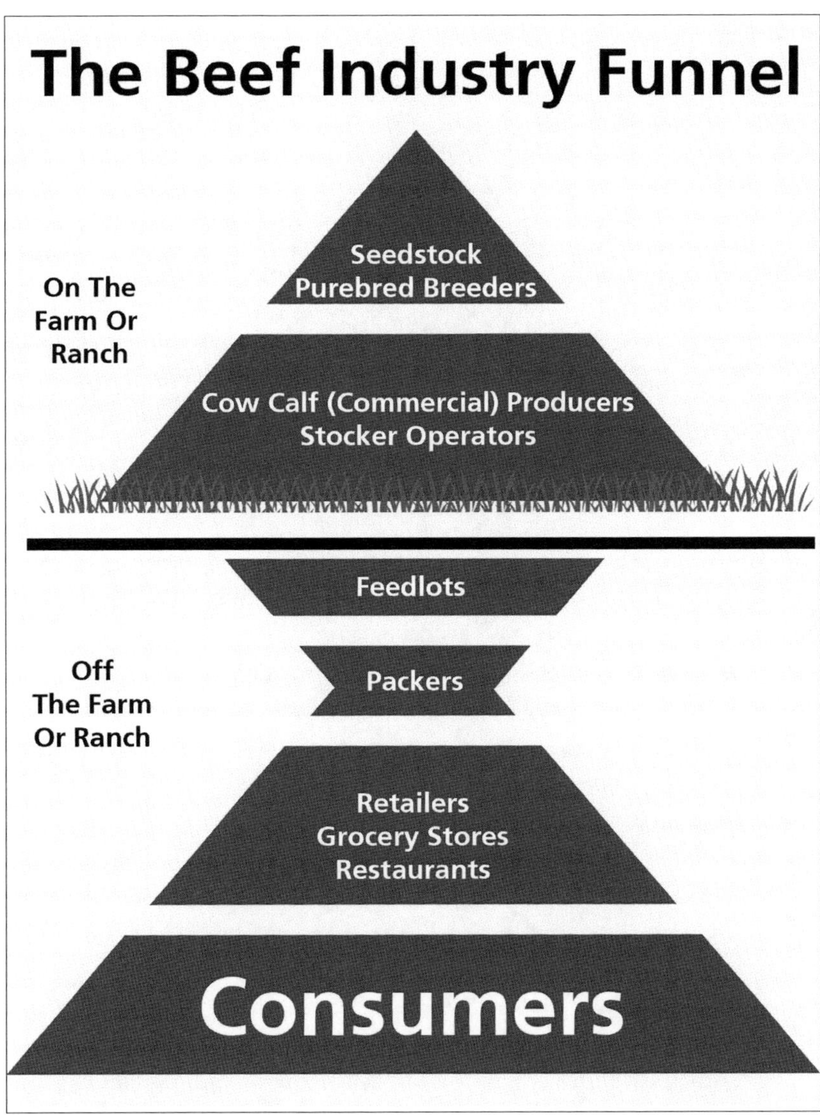

Promotional efforts can help increase the price you sell your livestock for on the farm or ranch. Once your animal leaves the farm, producers have little to no control over the price they receive for their product.

The fascination factor is a phenomenon that describes the special connection between human and livestock. Photo: Judy Jacobson

because of the influence of three key factors:

- **The purebred business is not just a livestock business; it is a 'people' business.** Public relations, customer service and overall like ability of breeders are all factors that play a role in marketing purebred products.
- **The price of purebred livestock is based on the quality of each individual animal.** When making buying decisions customers evaluate each animal on their own merit. Factors considered often include pedigree, winnings, potential for winnings, reputation of the breeder, individual animal performance, fertility, disposition, and performance of siblings or closely related animals. It is imperative that those involved in marketing purebred livestock hold a knowledge of their animals background, and be able to convey this to potential buyers.
- **"The Fascination Factor".** Neither science nor words can classify this phenomenon in livestock marketing. It is the remarkable feeling that certain animals give certain individuals. Perhaps it is an animal's color that draws a buyer's attention, or maybe the animal's personality. Some people may simply say that a certain animal 'has the look', or they can envision what a particular animal's genetics will be able to do for their operation. Experienced livestock producers also realize the fascination factor through envisioning the 'upside potential' of purchasing certain animals and their ability to make profits down the road through

progeny or genetic sales. Whatever it is it's hard to explain, but once the fascination factor is experienced it is unmistakable. It's addicting. It's mesmerizing. It draws buyers in and gives them the feeling that they must have a certain animal. This fascination, when felt by enough competitive buyers, has been known to cause a single animal to go from a $5,000 animal to a $50,000 prize in a split second.

Tools for Successful Purebred Livestock Breeders

There is no secret recipe for being successful in the purebred livestock business. However the following things are necessary tools that every successful breeder must have:

1. Availability of capital.
2. Availability to land, feed, and necessary facilities.
3. Quality livestock (both genetic and phenotypic).
4. Management and decision making skills.
5. A direction or 'game plan' for the breeding program.
6. Work ethic, patience, and perseverance.
7. A successful marketing program.

While these components are necessary, everything eventually comes down to marketing and sales. An operation could have outside sources of capital, land that is paid off, the best management, and the highest quality livestock, but without sales, a purebred livestock operation would fail.

When you look at the most successful businesses in any field they commonly share these characteristics: a quality product, brand name recognition; excellent service and customer relations.

For hundreds of years, livestock breeders have followed the mentality that improved production is their number one goal to success. Many producers subscribed to the thought that production improvements like increased weaning weights or higher calf crops were all it would take to be successful. However, beginning in the 2000s, many livestock producers saw a transition occur in their operations. Instead of being 100% production driven as in the past, livestock producers now had to concentrate on the combination of improving production and improving marketing.

Livestock producers today face many challenges, including limitations on land, water, and energy, as shown by the cattle supply chain graphic. While there are fewer and fewer producers, there are increasing individuals along the supply chain that provide services or charge fees that can deplete your profits. So today, just raising a great product isn't enough. Success in today's livestock business is both production-driven as well as market driven.

Marketing Options for Purebred Livestock

To get customers interested in your livestock, first of all, you have to have a desirable product that buyers will desire. Second, according to Bill Wilson, a longtime merchandiser and breeder of purebred Angus cattle, 'people want to buy from a strong firm that will stand behind their product with integrity."

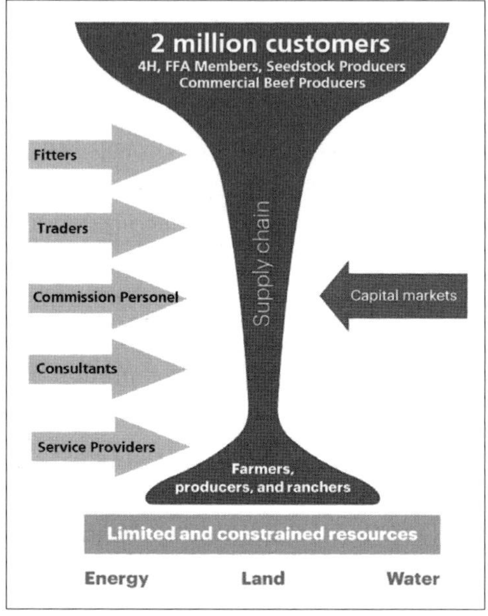

In the purebred livestock marketing process, one must always be on the offense. Remember the livestock you are selling are as much of an advertisement (good or bad) for your operation as any print advertisement or website. There is a buyer for every animal, and the marketing channels for purebred livestock usually occur through:

- Auction barns
- Private treaty sales
- Bid-off sales
- Internet marketing
- Consignment sales
- Annual production sales

Auction barns are the simplest marketing method, but leave very little room for premiums. This is a preferred option for some producers due to the convenience of being able to go to market at a scheduled time and local location, and then quickly receive a paycheck. It is very easy to haul animals to a sale barn on a predictable day of the week, unload them, and return that evening to pick up a check. When selling at an auction barn, one can have a fair idea of what the sale price will be, based on the current market report and the age and weight of the animals. The price that an animal sells for at the auction barn is really the base price, or true market value, of an animal. Anything above this price depends on the skill level of the marketer.

Private treaty sales are transactions that occur as private negotiations between the buyer and seller. These traditionally occur at the ranch, and sometimes at public events such as livestock shows. As technology becomes more advanced, some private treaty sales are actually completed as sight-unseen purchases by those who have seen photographs or videos emailed to them or posted online. However, a typical private treaty sale involves a personal visit to the ranch, in which the buyer meets one on one with the breeder. This gives buyers the opportunity to get a feel for the farm or ranch and look at the breeding program first hand. The breeder sets his price for each animal, and negotiates the deal with the buyer. Once the sale is made, the buyer pays the breeder and takes his animal home.

Private treaty sales have many advantages: they allow for casual interaction between buyer and seller, greater information exchange, personalized service, and help build lasting relationships. Sellers are able to deal one-on-one with prospective clients, get to know their goals, and help them find the best product to fit their needs. Private treaty sales also help build lasting relationships, establishing a repeat customer base.

The disadvantage of private treaty sales occurs when a producer has little knowledge of the animals' true value, and cannot confidently set prices. In some cases, the producer may actually never be able to determine the value of his livestock, and instead ask interested parties questions like "*Well, how much will you give me?* or "*How much do you think he's worth?*"

Producers who lack confidence in pricing their animals typically over-value, or under-value animals, thus making it difficult to complete the sale in a manner that is advantageous to both parties. When setting the price in private treaty negotiations, the seller should calculate his selling price based on the value of the animal as well as his costs, inputs, and any additional broker commissions to be paid. Despite a producers best efforts to price his cattle, it is easy to know when you have set your prices too high: you simply won't have any private treaty buyers.

Time commitments may be considered another disadvantage. Some prospective clients may visit your ranch several times before actually making

The Art of Private Treaty Sale Negotiations: A Tale of Two Brahman Breeders

Amongst Brahman breeders, there is a famous story that describes an actual private treaty transaction between two legendary breeders in south Texas in the 1960s. The story goes like this...

Mrs. Armel Baker was one of the early historic Brahman breeders, and one of south Texas' most wealthy landowners and cattlewomen. On one occasion, she had travelled to the Garcia Brothers Ranch in south Texas (almost to the Mexico border) in search of a Brahman bull. She had her chauffeur bring her cattle trailer along - as she was prepared to make a purchase. However, she had the chauffeur park the trailer in a deserted lot in the town so Mr. Eligio Garcia did not see the trailer and think she was overly interested in a bull.

After a full day of negotiations and trading between Mrs. Baker and Mr. Garcia, the sale was complete. As a tradition, the two breeders returned to the ranch hacienda to celebrate the business transaction with a cocktail.

As they approached the house, Mrs. Baker stopped Mr. Garcia and said that she had something important to tell him before they went inside. In a sassy tone, she said, "My dear Mr. Garcia, I wanted to let you know that I was prepared to pay A LOT more than what I did for your bull!"

And without missing a beat, Mr. Garcia said, "But Mrs. Baker, I was prepared to sell him for a whole lot less than what you paid."

a purchase, or may never make a purchase. Additionally, if offering private treaty sales, producers should be prepared to expect weekend visits, after hours visits, and clients who wish to come look at your sale animals over holidays.

The 'tire kicker' syndrome is another disadvantage to marketing your livestock through private treaty methods. Tire kickers are those who never really were interested in actually purchasing an animal, but who still make inquiries, even visits to the farm, to see your livestock. They may be looking for information, looking to scout out the competitions prices, or may simply be using your operation as a nice way to get out of town and spend a wonderful afternoon on the ranch. While you never know when a tire kicker might turn into a real client, they can drain hours of your time.

Private treaty pasture sales are a common variation of the private treaty sales method in the club calf sector. In a private treaty pasture sale producers specify a date and time that cattle will be priced and available for purchase. This creates somewhat of an open house event, attracting potential customers to the farm or ranch on a given date that can be prepared for in advance. When potential customers arrive, cattle are priced on a first-come, first-served basis and available to be purchased and taken home as soon as the transaction is complete. The sale date is only used as a method of attracting buyers and focusing a producer's sale season into a tighter time frame, compared to year round private treaty sales.

Here are additional tips for those using private treaty marketing:
- Do your homework before the sale. Livestock should be in good condition and meet all health requirements so that they may be transported to any state depending on the buyer.
- Sort your livestock into groups that are easily viewed by clients (i.e. by sex, by age, by price range). In general, the more pens the better. If you have fifty bulls in one pen, there will be one "best bull" in the pen. However if you have five pens of ten bulls, you will now have five "best bulls" of the group.
- Sort your keeper animals and any sold animals from your sale offering and in another location on the farm or ranch so that they are out of view. There is nothing worse than a potential client asking about a certain animal only to hear that it isn't' for sale, or is already sold.
- Have your livestock in good condition, with each animal clearly

identified. Utilize ear tags if possible to identify birth dates, sire, dam, and other information.

- Let your livestock first speak for themselves. Don't create a high pressure environment for buyers. Listen to your buyers, and help them find the best fit for their needs.

- Be honest about the attributes and strengths of your animals. If a prospective client asks what is wrong with a certain animal, tell them. It is better for your buyer to hear it from you before making the purchase rather than hearing it from every single person who looks at the animal in the future.

- Don't try to talk buyers into a purchase, because if anything goes wrong or the animal does not perform as the buyer wishes, he will feel as if the seller 'tricked him into it' or coaxed him into buying something he did not want.

- Accurately represent your product. Be truthful about pedigrees and birth dates. Have all animals registered and pedigrees available so that buyers can see the bloodlines.

- Be hospitable. Always be welcoming and friendly to each and every person who inquires about your program. Treat every client as if they might be your biggest customer, no matter if they are buying a $1,000 animal or a $10,000 animal.

- If you have a buyer who is undecided on a purchase, you can extend him an option to buy for a designated time period. This allows him a little time to consider the offer without you selling the animal to someone else. A typical option might be something like, "You have the option to buy this bull for one week. If anyone comes to the ranch and wants to buy this bull, I will check with you first to give you first right of refusal."

- Offer guarantees and stand behind your product.

A pasture sale bid-off is an auction style format that is not nearly as labor, time or financially intensive as hosting a live production sale, yet still allows for competitive bidding between potential buyers. In a pasture sale bid-off, also referred to as a telephone auction, animals are available for viewing for a period of anywhere from one month to two weeks, along with a designated date and time of when the bidding period will begin and end. During that time, potential clients can visit the farm or ranch at their leisure. Animals are all started at a base price and bids are usually placed in increments of $100 to $500 from the base price. As interested buyers come by to view the offering, they may place a bid on the animal at the next increment level. At the conclusion of the bidding period, the producer generally begins a telephone bidding process by calling each person who previously bid on a particular animal, giving them the opportunity to place another bid, or decline. This process is repeated over and over until a final high bidder is reached. At that time, the producer moves to the next animal in the sale, and repeats the process again. The telephone bid process can be very time consuming, but overall is less time consuming than hosting an actual live auction.

Internet marketing has become a favored marketing method within the purebred livestock industry, due to convenience and affordability. Internet sales first arrived on the livestock marketing scene in 2005, and has experienced tremendous growth since day one. In 2010, DV Auction, one of the nations leading internet sale firms conducted 220 online sales. In 2013, they conducted 650 online sales. The DV Auction data alone shows triple the growth in online sales from 2010 to 2013.

If a producer wants to utilize the internet in his livestock marketing efforts, he has three options. The first, and most basic option, is to simply feature your sales listings online either through your own website or through a directory service. Popular directory services include www.cattlerange.com, breed association sites, or even industry-specific message boards such as www.steerplanet.com.

The second option is to host an internet sale through a professional Internet sales firm. This type of method typically allows for eBay type bidding in which the sale is open for a specified time frame and interested bidders are allowed to place their bids as a sale countdown occurs. When the sale closes, the highest bidder is the winner of the lot.

Finally, the most intensive type of internet marketing option is when a ranch hosts a traditional "big tent" production sale and offers an online bidding component to allow for greater accessibility of bidders. In this case, the farm or ranch would host their normal production sale, hire an auctioneer, run the animals through the ring just as they normally would. However, the bidding would be open to both live bidders in the audience as well as remote bidders online. To use this type of service, a producer usually hires a professional firm such as DV Auction, Live Auctions, or Cattle In Motion to provide the internet bidding access.

How do you know if you are ready to host an Internet sale? Derrick Smith of Bulls Eye Ranch began using internet sales as their business's form of marketing in 2011. He felt like their criteria of knowing when they were ready to host an internet sale was when they had ten very high quality heifers to offer to the public. Smith stresses quality, citing that if you do not have enough cattle to offer 10 competitive show-quality animals, then private treaty may be a better method. In 2013, the average online sale had 17 lots. This option is a good option for those who do not have large enough numbers to host an annual production sale, but would like to move from private treaty marketing to some form of competitive bidding.

Web based auctions offer a happy medium between a private treaty pasture sale and a full fledged live auction. Sale animals are photographed and videoed to be posted online. The auction begins and ends at a designated date and time. Bidders must log in to a secure bidding system and place their bids via the internet, much like the eBay auction system. This type of sale is popular with producers because it is an inexpensive alternative to hosting a live auction, less labor intensive, and works well regardless of the number of lots being offered.

Here are some additional tips for those using internet sales:
- Schedule your sale date far enough in advance to do enough publicity to promote your event.
- Utilize at least three email blasts to promote the event (one when the sale information is posted, one when the sale opens, one on the last day of the sale).
- Assign one person from your team to be the go-to contact for those viewers having technical difficulty.
- Do not 'run' your buyers by placing false bids. Buyers can get a feel for this problem online, and are easily turned off if they receive notifications that they have been "outbid" 10 minutes after every time they place a bid.
- Remind your clients that due to differences in Internet connections, it is always a good idea to get your bids in early. Never wait until there are 5 seconds left on the lot to place your bid. While most systems have safeguards in place to give ample time for each bidder to close out their

bids, it is best to get your bids in early.

- Follow up with your buyers within 1 hour after the sale closes. A quick phone or email will help your buyers know what to expect as the next step. At this time you can discuss payment, delivery, and other necessary information.

When choosing a company to work with on your Internet sale, carefully evaluate each company's list of services, features, and number of registered bidders or subscribers. While there are several standard available systems for online auctions used by livestock Internet auction providers, each have their own unique features, advantages, and disadvantages.

Consignment sales are live auctions that allow for several producers to

Consignment sales are very popular at large scale livestock events where several producers bring consignments to combine with other producers in a sale. By pooling the resources of multiple consignors, sale expenses are shared by the group and can be significantly less than the cost of one rancher hosting their own sale.

pool their resources and host one auction together. In consignment sales, sellers bring the product to the buyer - compared to private treaty sales where the buyer often comes to the seller. This is a common marketing method at events like major stock shows, beef expos, or for regional producer groups. Some state breed associations also offer consignment sales as a service to their

Preparation Helps Build Premiums at Sales

Hosting an annual production sale takes year round planning in order to be successful. Here are some key reminders for having your livestock at their utmost level of marketability for an auction.

Once your sale selections are made, sort these animals from the herd and begin increasing their nutrition plane. Mark Cowan, livestock marketing professional, advises, "There is a difference between being in good pasture shape and sale shape. Recognize the difference and have them in 'sale' shape."

For beef cattle, bulls should be fed to maximize genetic potential, but still maintain soundness and fertility. Females need to be in body condition score of 6 or higher. Nursing females, especially first calf heifers, may require several weeks to be sale ready, especially if they have been allowed to be nursed down, so allow plenty of time.

Disposition is also important and can build premiums. "Halter broke" and gentle are two descriptive terms that can be used in sale terms as an added value. While it certainly isn't a requirement that all sale cattle be halter broke, it certainly is a bonus when offering show cattle.

members. When participating in a consignment sale, producers usually pay a nomination fee to help cover such sale expenses as advertising and catalog production. It is customary for producers to also pay a sales commission to the organization hosting the event (i.e. 10% of total sale price).

Consignment sales are appealing because they represent a collective sales effort between all of the parties participating in the sale. Small breeders, especially, can take advantage of consignment sales because they can bring in larger numbers of animals and a variety of genetics, instead of single sourced genetics from one operation. There are few breeders who have the quantity, quality, and capital to hold their very own production sale. But by joining forces with other breeders either within your local area or within a breed, you can team up to do a superior job of marketing. Consignment sales offer the advantages of targeted locations, joint promotional efforts, trained sale managers, and more publicity to the potential buying audience.

Suggestions for successful consignment sale participation:

Barber Ranch hosts an excellent ranch production sale. Each animal is clearly identified with a sale ear tag featuring their lot number. Cattle are sorted into convenient pens for ease of viewing. Members of the Barber family, like Terri (pictured left) are readily available to interact with buyers and answer any questions.

- Offer your highest quality animals. This is a great opportunity for breeders to showcase their program to people who may not ever visit the sellers ranch. Never use a consignment sale to get rid of culls.
- Advertise your consignments in the sale prior to the event.
- Contact your established client base to let them know you will be putting animals in the sale, and invite them to the event either though a personal letter, email, phone call, or all of the above.
- Submit a good photograph along with a well written footnote to be

Express Ranches host a first-class sale. Pictured above, their classy sale auction block adorned with banners of Express show winners from previous years. Express also offers internet bidding and phone bidding options to make it easy to do business with them.

included in the sale catalog.

- Obtain copies of the sale catalog prior to the event and mail them to your selected clients.
- Arrive with your animals early enough before the sale that they become acclimated to the sale environment.
- When the sale day arrives, make sure each animal is in good condition, clean, well fed, and properly identified.
- Hang an attractive sign with your name and contact information in your pen area.
- Have any necessary paperwork in order prior to the sale. If selling registered animals, have the papers ready to be transferred as soon as payment clears. Have health papers available for interstate shipment. If selling pregnant females, offer a certificate from your veterinarian showing the female to be examined safe in calf. If selling a bull, provide a copy of the bulls fertility examination.
- Be around your animals before the sale to answer questions from prospective buyers.

Production sales are very common for farms and ranches desiring to market their total production for the year all at once. Typically an operation would host an annual production sale at the same time each year (i.e. Labor Day weekend), allowing the event to become one that fellow breeders anticipate and mark on their calendar every year.

Production sales are extremely time intensive and require year-round planning to maintain the annual date. Breeding decisions, culling decisions, herd management and promotion all revolve around this one annual event. Mark Cowan, a professional livestock marketer, estimates that the success of a production sale is based upon a minimum of the past three years management of the operation. Auctions are a very popular means of marketing cattle but can be disastrous without lots of planning, strategy, hard work, and effort. The days of putting up a tent, running a few ads, and expecting a successful sale with little or no effort just do not exist anymore. And ultimately, in many cases, the ranch's gross operating income for one year all boils down to what happens on one single day: SALE DAY.

Many producers who host an annual live auction also choose to hire an internet broadcasting firm, such as DV Auction, Superior Livestock Productions, Cattle In Motion, or LiveAuctions.tv, to allow for real-time live viewing and bidding during the auction for clients who are unable to attend the event.

Should you have a public auction? In order to have a good sale, you must first have top quality, gentle cattle that are in good shape with popular pedigrees. Breeders who have no trouble selling their product privately are the ones that can benefit the most out of a public auction, because they may never truly fully realize the value of their livestock unless sold through a competitive bidding process. For example, a well known Florida ranch was selling all of their bulls privately at their farm year round for an average price of $2,500. In the late 2000s their demand grew so great that they had too many people on a waiting list to get first pick of their bulls. The ranch then decided to have a public auction, and over the past three years, the bulls have averaged between $3000 to $5000, with a high seller in 2014 of $10,000. Without a competitive bidding process, that $10,000 bull probably would have sold in the $2,500 pen just like all his other contemporaries.

The most reputable breeders within specific breeds or categories should also consider holding public auctions. Smaller and lesser known breeders look to the larger, more well known operations to set the market value for others to utilize. Many people that may be interested in a breed read sale reports to determine whether they feel demand for a product is increasing, stable, or declining. Prices from public auctions serve to either encourage or discourage people investing in your segment of the industry.

If you host a sale at your ranch, producers should make every effort to see that the people attending the event are welcomed in a friendly manner, treated with respect and hospitality, and provided all necessary information to help them make their buying decisions. An upbeat, positive sale atmosphere will not only encourage prospective buyers to invest in your program, but also to encourage their friends and associates to come and visit you.

Special considerations should be given to displaying animals at sales, where the goal is to display your product in a manner that will make it easy and enticing to prospective buyers. The animals in the sale should be gathered in one location in a convenient manner for clients to visually evaluate each animal. All sale animals should be clearly identified with the lot number, whether by ear tag, body paint, etc.

Construct your pens or display in a manner that is most visually appealing to the animal. For example, cattle look better when displayed on grass compared to a dirt lot. Fresh, clean shavings also provide a professional bedding for displays. Having reference sires and dams on display can be advantageous, as well, to allow viewers the opportunity to view the sires and dams of sale animals.

Buyers appreciate pen diagrams of production sale layouts. Breeders should develop
 a handout that contains a pen map, easily depicting where sale animals are located. This will help buyers to locate animals they are interested in for visual inspection.

Hospitality is an important aspect of creating a welcoming and professional appearance. Seasonal decorations, like fall pumpkins or spring flowers, can be added around the sale ring for a nice touch. A professionally designed sign featuring the company logo, should hang at the auction block. Road signage should be used to direct visitors to the sale facility. Even signs as simple as small ground-staked arrows, including the company or sale name, pointing in the direction of the sale location are extremely helpful.

Sale management firms, experienced ringmen, knowledgeable auctioneers and livestock brokers can help tremendously in taking the marketing burden

Experienced ringmen and a professional auctioneer are some of the most important people that contribute to a sale's success. Upbeat ringmen help create an energetic atmosphere and set the tone of excitement. A professional and knowledgeable auctioneer can often make or break a sale. Pictured above, some of the best in the business: Kent Jaecke, Delvin Helderman, Chisolm Kinder and auctioneer Steve Bonham.

off of the producer, especially those who host annual production sales. Sales management firms come in and handle a variety of services from catalog production, to procuring the auctioneer, to facility set up and take down. For a percentage of the sale proceeds sale managers serve as the go-to-guy (or gal) for the sale and offer a great deal of marketing expertise and service to the producer. If hiring a sale manager, make sure to state specific responsibilities clearly, so that both the seller and sale manager know what each party is responsible to complete.

Livestock brokers are more commonly part of private treaty negotiations. The livestock broker helps build initial contacts with prospective clients, for a commission based fee that typically ranges from 5 percent to 10 percent of the gross sale. Livestock brokers are extremely common in international livestock transactions, in which the broker provides services such as translation, working with the government officials to make sure the livestock meet health requirements, securing transportation and shipping crates, and other mediation services between the buyer and seller. If you are working with a livestock broker, it is important to discuss the commission terms up front, and also describe these in writing so that both the buyer, the seller, and the broker are clear on the responsibility of each party.

Regardless of the marketing method you choose, honesty and integrity are crucial to success in livestock marketing. Do not misrepresent your product. Never make statements you cannot back up, such as guaranteeing a certain animal to be a show winner or high seller. If your animals have a fault, be honest about it. Buyers will appreciate your honest and value your integrity.

At some point along the way, most individuals will be faced with ethical decisions about marketing their cattle. Should you "run your buyers" in an auction sale? Should you allow for animals to be sold prior to the sale and still run through the auction? Should you misrepresent your prices or sale averages in order to make your operation look better? When asked to define integrity, some would say that integrity is what one does when no one else is looking. The principles of fairness, moral integrity, and honesty should all be used when marketing livestock. It is better to maintain your reputation than participate in a shady livestock business transaction.

A Word of Advice to Buyers

The majority of this book is devoted towards breeders and sellers of purebred livestock, in which we offer tips on how to more effectively promote your cattle. However, we feel it is equally important to offer tips to buyers... since most livestock producers will play the role of both buyer and seller throughout their tenure in the livestock business. Guerra recommends all new breed enthusiasts visit at least five different operations prior to investing in their first purebred livestock purchase. Personal visits are best, however if that is not possible due to time or financial restraints, take the time to visit ranch websites before making a purchase. Factors to consider include:
- Number of years in business and integrity of the breeder
- Quantity, type, and quality of livestock offered
- Awards earned
- Program philosophy (management, breeding, etc)
- Major bloodlines utilized

After researching, and preferably visiting the operations, buyers will have a much clearer idea of the type of animal they would like to own. At that time, buyers should set goals for their operation. Discuss how big or small you would like to be. Do you want to own an interest in one high power donor cow, or own 200 head of breeding age females, or both? Do you want to manage the livestock personally at your own farm or ranch, or be an absentee owner? Do you plan to show competitively?

These questions should be answered early in the process so that the buyer can have a clear direction of their goals. Purchasing and owning purebred livestock requires a significant investment in time, money, efforts, and more. If you don't know where you are going, you will have no way to get there.

Avoid dealing with unscrupulous peddlers, traders, and scalpers. When buying purebred livestock, get to know the breeder or seller and get as much information as possible on the past history of the animal and breeder before making a purchase. If you know little about pedigrees, consult someone who does. If buying livestock at sales, buy at reputable sales and purchase animals consigned by reputable breeders.

There is no substitute for high quality livestock. Cheap animals are the most expensive livestock you can buy...as they are an expense and not an investment. You can spend a fortune on cheap animals and spend a lifetime trying to breed them up, but in the end, only quality reproduces quality.

Buyers should consider the following advice when making any purchases in the purebred livestock business.

1. Buy the best animal you can afford.

2. Breed the best sire you can afford to the best cow you can afford.

Owning quality livestock is a way to continually build and grow agricultural assets over time. As time goes by, the difference between great livestock to mediocre livestock continually becomes larger and larger.

SOURCES:

Buying a Bull? Remember these Basic Guidelines, James E. Pace, Gainesville, Florida, article published 1972.

Cattle Ventures Lead to Business Profit, Teresa M. Spivey. Published in The Drovers Journal, Roll Call. 1975.

Marketing Makes The Difference, Brent Thiel. Charolais Journal, October 1986.

Sale Day Preparation, by Merridee Wells. Simbrah World, Fall 2010. Merchandising Bulls at Private Treaty, Martha Hollida. Brangus Journal, September 1982.

Personal interviews: Brad Fahrmeier, Carlos Guerra, Tim Lockhart, Derrick Smith, Jim Williams

CHAPTER 2:

Components Used in Livestock Promotions

Just as any good reporter begins a story with the basics of who, what, where, when and why, a rancher should address these same basic questions before beginning an advertising campaign.

Who?

This book is primarily directed at ranchers, farmers, and purebred livestock breeders and owners who are placing advertisements to promote their animals, farm, events or a combination of all of these things. Typically advertisements will be placed by ranchers, breeders, owners, sales managers, and other individuals who have a product to promote. However, this book is also designed to be a tool for students aspiring to one day become livestock marketers or graphic designers.

The first form of advertising was done by word of mouth, and today, it still remains one of the most effective means of advertising and promotion. Personal communication such as telephone calls, a letter, an email, or most effectively a personal visit, help build relationships and strengthen a marketing program.

What?

In the purebred livestock sector, most advertisements revolve around one crucial element: livestock. In this book we will discuss advertising for cattle, sheep, goats, and hogs, occasionally using examples from the equine industry. Within these broad categories, one will find several subcategories of livestock advertised:

- Seedstock being promoted for breeding value (sires, breeding females)
- Young animals being promoted for the purpose of show competition
- Genetic material including semen, embryos, and flushes

Where?

Where do you advertise? Do you consider advertising an investment or an expense? Currently there are thousands of media outlets competing for the livestock producers' advertising dollar. For this reason, owners should carefully evaluate each prospective media outlet to assess the cost vs. benefit of each media source. A few of the current media outlets being used by ranchers include:

- Printed publications (breed magazines, newspapers)
- Direct mail pieces
- Sale catalogs
- Digital marketing outlets including websites, social networks, video
- Outdoor displays and advertising

Research done by numerous livestock marketers show that the average buyer is willing to travel 100 miles or less from home to make a purchase. However, in the purebred livestock business, this location can be extended nationwide as many producers are willing to travel hundreds, if not thousands of miles in search of the great one.

With this in mind, producers should consider advertising on both a regional and national level. Use popular, well accepted advertising outlets to cover your marketing area.

When?

The modern livestock marketer will participate in on-going advertising and public relations campaigns to achieve their desired results. Within the span of a year a producer's advertising will focus on different events. A typical advertising calendar for a purebred livestock producer may include the following:

- Winter: Promotion of sire prospects, Denver display bulls and results of winter major stock show results.
- Spring: Promotion of spring bull sales and continued emphasis on semen sales.
- Summer: Early promotion of fall production sales, State Fairs, and Junior Nationals.
- Fall: Heavy promotion of fall production sales, fall bull sales, fall breeding promotions, brief focus on fall major livestock shows, such as Kansas City or Louisville.

When you start advertising, don't expect results instantly. The best advertising and promotional plans involve constant, ongoing efforts. Often times breeders find it takes two to three advertisements before they start seeing results.

Why?

Ranchers and farmers have various reasons for placing paid advertisements in media outlets. In its most basic form, advertising is used to sell a product. However, the reasons are multi-faceted when livestock producers place advertising. Just a few of those reasons are as follows:

- To sell livestock.
- To create brand name awareness and recognition.
- To promote the positive goodwill efforts of a ranch or farm.
- To create awareness and recognition for a specific animal (herd sire, show ring champion, donor female, etc).
- To inform the public about a business or product (announce a new purchase or accomplishment of the operation).
- To educate clients about the benefits of purchasing animals produced (performance information, production records, etc).

Now, more than ever, it is important to advertise. As competition amongst breeders and breeds becomes more intense, every livestock producer needs to take an aggressive approach to advertising and promotion, and ultimately marketing their product.

Livestock advertising, even in its most elementary forms, has been used for centuries. One might venture to speculate that early drawings on cave walls were a form of advertising the superior livestock of prehistoric times, when animals were used for draft and work. When the printing press was invented in the 1400s, it became possible to print large number of copies of advertisements that could be easily distributed in the marketplace.

Early livestock advertising designers had it rough. There were no computers or scanners, no InDesign or Photoshop, and no stock photo services. Therefore, livestock advertising from the early 20th century relied on the basic principles of readability, clarity and text based information. These ads were primarily designed on white backgrounds, using large bold black type, perhaps a photo, and persuasive ad copy that often explained the benefits of using a particular ranch or breed. In the 1940's the use of spot-color (one color) in livestock advertising began to emerge, often used minimally on magazine covers only. As the use of spot-color continued to increase, livestock magazines eventually began using the four-color process as early as the 1960's. While the four-color process cost more in advertising rates, it became extremely popular and most producers began seeing full-color advertisements filling the pages of every livestock publication. By the early 2010s, the technology and availability of full-color printing led many publications to printing 100% in full-color as their standard service, without charging a premium for the color placement.

Livestock advertising has greatly evolved through the years to arrive at the predominantly full-color, high gloss formats of magazines, advertisements and sale catalogs seen today. The addition of the internet and social media also revolutionized livestock merchandising drastically since 2005. Today's smart livestock merchandisers use a combination of digital and print advertising to offer the full spectrum of exposure for their businesses. They utilize print media for tried-and-true promotion during key times of the year to help publicize sales and events. Then they use the internet and social media for their go-to outlet to provide timely news and updates for an instantaneous delivery method.

How Do I Decide What Works for Me?

Regardless of the size or type of your livestock operation, there are thousands of tools that livestock producers may use to help create their own unique brand identity and promote their livestock. The tools mentioned in this chapter are the framework for establishing a marketing and promotion plan. Ultimately, they are used to help potential clients learn about your product, begin to recognize your business, and be able to distinguish it from others. While it will not be necessary for all operations to use every method that follows, these outlets are the most common and successful promotional tools used by the best livestock marketers.

Basic Components
Colors
Name
Logo
Tagline or Slogan

Printed Advertising Components
Brochures
Business Cards
Catalogs
Holiday Cards
Letterhead & Envelopes
Magazine & Newspaper ads
Newsletters
Postcards

Outdoor Advertising
Farm Signs
Billboards
Pedigree Signs
Ranch Banners
Promotional Displays
Show Animal Exhibition

Broadcast Advertising
Internet Broadcasts
Radio Advertising
Television Advertising

Web-Based Advertising
Blogs
Email Blasts
Email Addresses
QR Codes
Social Media
Websites
Web Banner Advertising
Web Links

Other Advertising
Apparel
Photography
Promotional Give-A-Ways
Sponsorships & Contributions
Text Messages

The Basics
Name

Creating a name for your operation is the first and perhaps most important step in beginning any business promotional plan. Your business name is the primary unique way that consumers will identify you in the marketplace and distinguish you from others.

For those preferring a more traditional business name, your task is easy. Many choose to incorporate their own name combined with the words farm, ranch, cattle, or a specific breed. Examples of traditional names are GKB Cattle (standing for Gary & Kathy Buchholz), Dismukes Ranch, or Sullivan Farms.

V8 Ranch is named after the V-8 engine, because their original ranch owner also owned a car dealership and he favored the V-8 engine over other cars. Logo by Martha Hollida Garrett.

Using a ranch brand in your name is also common, such as Rocking B Ranch or 6666 Ranch. Simplicity and clarity are benefits of choosing a more traditional name. Keep in mind however common last names are often in current use, thus limiting name selection.

If you prefer a more nontraditional name, consider creating something unique that tells a story about your identity. Tennessee River Music is a registered Hereford operation in Alabama, owned by Randy Owen of the famed country music band Alabama, and his family. You may recognize the ranch name from the lyrics of one of the band's popular songs, "Tennessee river and a mountain man, we get together anytime we can........"

After you have chosen a few potential business names, it's a good idea to search the Internet to determine if your name is already in use by someone else, and if your name may be available in the form of a .com domain name. While creating a business name is at no cost to the producer, it is a wise decision to trademark or copyright the business name, which may result in additional fees.

Logo

Is a logo really that important? Two words: golden arches. This icon is a timeless example of how a shape can create instant brand recognition and identity by one of the most successful corporate advertising campaigns in

The GKB logo incorporates the business name, ranch brand, and red colors.

The Pembrook Cattle Company logo utilizes their herd prefix (PCC) as the main visual element, in a clear type style that is easy to recognize.

history. Just as leading international brands incorporate a logo into their marketing campaigns, livestock producers should do the same.

A logo is a specific type style, shape or graphic that attaches to a business an easily recognizable symbol. A good logo is one that can be utilized for lengthy amounts of time and provides a unique visual identity for a business. Logos come in a variety of shapes, colors, sizes and styles. For example, one company's logo may be as simple as a particular type style that displays the business name. The same type style is then utilized every time the logo mark is exhibited. More sophisticated logos include a graphic element, such as Nike's® swoosh or Apple Inc.'s ® apple, to add a visual element to the design.

If a graphic element is desired in a farm or ranch's logo, consider using the holding brand of the company. Many producers in the southern and western states utilize branding to identify their livestock. For these operations, their logo is already created, and most likely registered with an associated state agency.

As many individuals have access to graphic design programs such as Photoshop, the do-it-yourself option is becoming increasingly popular. Livestock breed associations may offer their members free use of standard graphics such as the standard Angus bull or a pig outline that can be utilized. It is imperative that whatever program is used to create a logo allows for a compatibility with different types of printers or software platforms. Typically, logos will be saved and named in either a JPG, PDF, PSD, TIFF or EPS format. Advanced graphic designers will often utilize vector logo formats created in Adobe Illustrator or Corel Draw. Logos should be created at high resolution (at least 300 dpi) which allows for large scale reproduction.

If hiring a professional to create a logo, it is important to present the designer with clear direction on the type and style of logo preferred. For instance, a western look may be the desired style for some, while a very minimal look is often more appealing to others. The designer will need to know if only specific colors will be utilized, or if the preference will be to adjust the logo using several different colors in the future. Whether or not a shape is to be incorporated into the logo, or only a type style used, should also be conveyed. Discussing and presenting these preferences prior to any actual design work will save time, efforts and often money throughout the design process. A professionally constructed logo can cost anywhere from $100 to $1,000 for a custom creation.

Depending on the intricacy of a logo, business owners may want to consider copyrighting or trade marking the design. It is also very crucial to avoid any type of copyright or trademark infringement when creating a logo. Avoid imitating another company's logotype, shapes or graphic elements, unless written permission has been obtained from the owner granting license to use the images. Never save an image from a website to use for design purposes. When in doubt, it is better to err on the side of caution; it is best not to use an image that may be too closely related to that of another.

Once a logo is created, use it everywhere: on caps, apparel, tack, stall signs, farm signs, printed materials, company website, business cards and anywhere else imagined. If you have a livestock trailer, it is recommended to hire a professional trailer graphics company to paint or create customized lettering including the logo for placement on the equipment. Remember, the goal is to continue placing the logo in front of potential clients. The more the logo is seen, the more recognition consumers will begin to build for the brand.

Tagline or Slogan

A tagline or slogan is a set of words that are used to describe a business or the services offered. Typically, slogans are used for temporary advertising campaigns and taglines are a permanent set of words describing the business.

Upon hearing the words 'eat fresh', the trademarked tagline of Subway restaurants comes to the mind of many consumers. Likewise, most everyone can identify the product that 'melts in your mouth, not in your hands' as Mars Incorporated's M&Ms®. Usually, slogans and taglines range from one to seven words. When creating a tagline, longevity is the goal. Choose a set of words that are timeless and can be used for years to come.

Begin the process by jotting down a few key phrases that describe the business, the product, important values upheld and any distinguishing characteristics. These phrases can often help ignite an idea that can eventually become a slogan.

For traditional operations, a tagline can be as simple as describing the service or product offered. Examples of this would be something like Johnson Family Farms: Family-owned and operated since 1988; or J.D. Hudgins, Inc.: Beef-Type American Brahmans. These phrases tell the consumer about the businesses and their values.

Non-traditional taglines can also be used to convey the overall goals or values of a business. For example, Express Ranches uses the phrase "The One to Tie To", thus conveying thoughts of stability, loyalty and dependability.

The cost of creating a slogan or tagline is free, however it is recommended that businesses using a non-traditional slogan make the financial investment to trademark their creation through the U.S. Patent and Trademark office. This can be done by the owner, though it is somewhat of an ominous task, or with the help of a trademark lawyer.

Colors

If one were to walk through a drug store and spot a bright pink liquid in a bottle, chances are they would instantly think of Pepto Bismol®. The same principles apply to agriculture; if a farmer sees two tractors, one green and one red, chances are they will quickly recognize the tractor manufacturer.

Color psychology can also play a role in choosing an operation's signature colors. Leatrice Eiseman, color expert, explains the importance of communicating with color by describing the feelings associated with certain tones.

Red signifies love, retail sales, anger, and also emergency. It conjures feelings of energy and is attention getting.

Orange is a fun and playful color that is used in youthful themes. It is popular among agricultural circles because of it's connections to agricultural universities like Oklahoma State University.

Yellow is a warm and cheerful color that also implies energy, freshness, and optimism. Yellow is a color that can be hard to read, and must be used carefully.

Green is a soothing, natural color that represents the environment, money, and even jealousy or envy.

Blue tones produce a calm and stable feeling, often creating feelings of trust and power. Blue is widely used in livestock circles for the association with quality, or a blue ribbon.

Purple is also a very popular color choice for livestock producers because it is the color of champions. Purple also implies royalty and is a very luxurious color.

Black, golds and silvers are commonly selected by agricultural producers because of their association with luxury, elegance, power and strength. It is also used in the cattle business frequently because many animals are black-hided, thus these tones match the animals very well.

White and beige are also colors that can be utilized in your color selection process to imply professionalism, crispness, and class.

Once specific colors are chosen, they should be exhibited everywhere: on equipment, business apparel, caps, tack signs and most certainly in all printed advertising. When choosing an operation's signature colors, select colors substantially different from those of any close business competitors; this will help consumers begin to distinguish the business from others.

Printed Advertising Components
Brochures

Brochures are often used as a bridge between a one-page advertisement and a catalog. Brochures are great for special promotions and general information.

There are several types and sizes of brochures with the most common being the 8.5x11 inch tri-fold format. This project often includes a standard set up of the company logo on the front panel, company contact information on the back panel, and the bulk of the information and photographs contained on the inside panels. Brochures like these are easy to create and often inexpensive to print. They also fit easily into a standard business envelope, making them easy to mail.

Business Cards

A business card is one of the most fundamental components of an advertising program, and often one of the most affordable. Business cards should be handed out at every opportunity, even tacked on bulletin boards at every feed store, auction barn and vet clinic in the surrounding area.

Business cards are typically 2x3.5 inches and printed on a heavy weight paper such as a gloss or matte finish card stock. At minimum, a business card should include the business name, logo, slogan, and contact name and information. Include a mailing address, phone and fax numbers, email address and website address. Include as much information as possible, making it as easy as possible for potential clients to contact the business. Of course, a business card should be designed using the company colors.

Circle T Cattle Co.'s business cards feature a double sided design. The front of the card includes the ranch name and slogan. The back side is personalized for individual staff members and lists contact information, and staff responsibilities.

A professionally designed, upscale business card is always more impressive than one created on a home computer. Business cards can be upgraded to include UV coating, special shapes such as rounded corners or oval designs, and can even be double sided.

There are many online vendors that allow individuals to create their own business cards at very affordable prices. These websites have easy to use templates and allow for the uploading of logo images and the like.

Creativity can make a business card stand out. Including a reference beef gestation table on the back, printing a map to the headquarters, or even using a photo of a prized animal produced by a farm or ranch on the front of the card and then placing the contact information on the back can make an impression on customers.

Catalogs

Catalogs are perhaps one of the most time consuming and largest advertising projects that livestock marketers will work on as part of an advertising campaign. A catalog is considered to be any printed material that is greater than four pages and is generally bound by saddle stitching, staples or perfect binding. Catalogs printed by a professional printer and professionally bound must be designed in page increments of four (i.e. four, eight, twelve, sixteen pages). Typical uses for this medium includes sale catalogs, sire catalogs and donor catalogs.

Circle A Angus Ranch Sale Catalog

Advance planning is required for catalog development in order to write content, obtain photos, print and mail.

Holiday Cards

Custom holiday cards are a personal touch that helps to create friendship and personal relationships with clients and colleagues in any industry. These small gestures are a great way to tell customers how much they are appreciated, and provide well wishes during the holiday season. The most noticeable holiday cards include some type of photograph, whether it be of the owner's family or staff, or of the headquarters or product of the farm or ranch. If writing a personal message, avoid excessive bragging and remember to include major highlights of the year, humor and a wish for good will.

Letterhead & Envelopes

Business correspondence such as letterhead and envelopes help aid in the professional image of a business. Custom letterhead should be used for all business correspondence through mail and fax. Typical letterhead for agriculturalists includes the business name, logo, ranch colors and contact information.

Letterhead can be created in a variety of formats. The simplest form is often done by using Microsoft Word® and inserting a logo, as a graphic, into the Word document. Business contact information can be inserted into the header or footer of the document as simple text. This type of letterhead, while not ideal, can be created at no additional cost to a producer. For a more upscale letterhead, heavy weight, watermarked bond paper with the logo and contact information pre-printed provides an extremely professional look.

Additionally, farms and ranches should invest in professional envelopes displaying logo and contact information in the return address area. Another very affordable option includes printing return address labels or purchasing a personalized return address stamp to help reinforce the business identity on every piece of material that leaves the office.

Magazine & Newspaper Ads

Placing advertisements in printed publications is a great avenue for getting a business name and its production in front of a large audience, specifically interested in a particular industry. There are hundreds, possibly thousands of potential periodicals an agri-business owner can utilize, with the most common being:

- Beef industry publications (*Beef, Drovers, Working Ranch*)
- Nationally based breed publications (*Angus Journal, Quarter Horse Journal, Seedstock Edge*)
- Publications segmented for certain segments of the industry (*The Showbox, Feedlot Magazine, The Showtimes*)
- Regional newspapers (*Weekly Livestock Reporter, Country World*)
- Regional or state publications (*The Cattlemen, Illinois Beef*)
- Local publications such as state breed association newsletters
- Local newspapers

Because of the costs associated with print advertising, careful and strategic planning should be conducted before beginning a printed advertising campaign, to ensure the maximum exposure and reader response. It is also recommended that ranches have a functioning website in place before placing print ads, so that there is an online presence to direct the reader to view.

Newsletters

Newsletters add a special touch to an advertising and public relations campaign. Typically a printed newsletter is somewhere between one to four

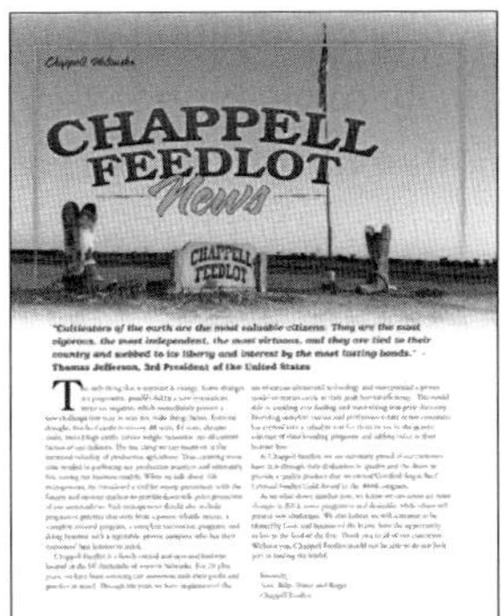

Chappell Feedlot publishes a newsletter with customer spotlights and educational articles.

pages, and sent out on regular frequency, whether quarterly or monthly. Effective newsletters include an update from the business owners or management, photographs, upcoming events and an informative news article of interest to the readers. For example, a newsletter sent by a bull stud may include an article concerning proper semen handling techniques.

The quantity and frequency of printed newsletters is decreasing due to the increasing costs of printing and postage. However, electronic newsletters are increasingly popular as they are inexpensive and can be distributed instantly. Electronic newsletters can be as simple as typing an update right into an email, or professionally designed and emailed as PDF attachments. Programs like Constant Contact® are popular for creating and sending electronic newsletters.

Postcards

Postcards are an affordable tool that can be used as handouts or as direct mail pieces. Postcards come in a variety of sizes and shapes, with the most common being the 4x6 inch or 5x7 inch formats. When choosing a card size and shape, it is important to consider postal regulations and charges. To qualify for mailing at the reduced postcard rate, pieces must fit the postal service parameters. The 4x6 card typically meets the postcard rate parameters, while the 5x7 card requires a first-class postage.

If a promotional postcard is simply to be used as a handout, many producers include a full size photo of an animal on the front with additional information on the back. This is a very common method for bull promotions.

When intended for mailing, remember the postal format requirements.

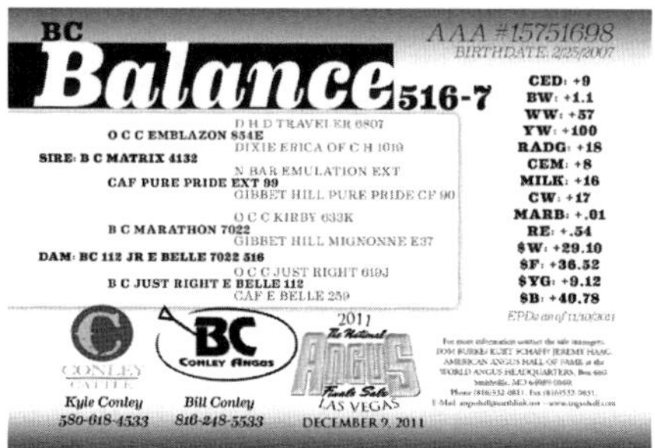

Sire cards typically feature a 5x7 photo on the front of the card and pedigree, performance, and contact information on the back. These are designed to be handouts, not self mailers.

The location of labels and stamps, as well as where the postal bar code will be placed, is important to consider during the design process.

Sales Tickets & Invoices

All businesses should have an official sales agreement or invoice that is used to create a record of all sale transactions. It is best if this form allows for the date and signatures by both parties and is printed in duplicate, which allows for the seller and the buyer to maintain a copy of the signed document. From a design standpoint, these forms are usually printed in black and white to reduce costs and should include the logo, name, and contact information of the business at the top. From a legal standpoint, it is recommended that all ranches have the basic terms and conditions of their sale printed on the document as well.

V8 Ranch sales tickets.

Agricultural lawyer Cari Rincker, (cari@rinckerlaw.com), is a great resource for creating sale terms and partnership agreements.

Outdoor Advertising Components
Farm Signs & Billboards

The farm sign is used to make it easy for visitors to locate your business and serves as a permanent storefront to every person who passes. A farm sign should be clearly visible from the road, making it easy for people to identify the entrance from several feet away. This sign should also display the business name and the product offered. Typical farm signs might include something like "Cates Farms - Polled Shorthorns" and may include a phone number or website address.

Though not as common as farm signs, billboards provide a promotional opportunity to producers owning land along major highways. The concept is basically the same as the traditional farm sign, yet reaches more people due to the size and scale of highway traffic.

Cates Farm uses a small sign and flower bed to create a welcoming entrance.

Promotional Displays & Animal Exhibition

For livestock producers in particular, attending livestock shows, fairs, sales and beef expos is the most common outdoor advertisement of their product: the animals. With this in mind, exhibition of livestock is one of the most important (and most expensive) forms of marketing for farms and ranches. It is imperative that livestock producers put their best foot forward when attending a livestock event where their animals will be on display. This is typically achieved by first and foremost: displaying the livestock at their utmost level of quality. Second, the ranch name should be prominently identified by signage. Finally, every animal that is on exhibition should be clearly identified with an individual sign displayed above their bed or stall.

Showing purebred livestock is one of the most expensive, but most beneficial types of advertising. Outdoor displays like shows provide an opportunity for large numbers of potential clients to see your product and compare it to your competition.

For larger shows, farms and ranches should consider setting up a small booth or seating area near their stalls or display to allow for clients to have a seat and visit. Drinks or snacks may be provided for hospitality, if facility regulations permit. Booth spaces can range from simple chairs and a flower arrangement to high-end upscale presentations. Remember the goal is to make a comfortable area for your clients, as well as a place to sit down and negotiate business deals if necessary.

Special displays, such as Denver bull promotions require a bit more planning and creativity. Traditionally, these bull owners get elaborate with these displays, as each is competing for the crowd's attention. Keep in mind that while bells and whistles create crowd hype, ultimately your potential clients want to see the animal clearly and be able to easily evaluate the quality of the individual animals in your display.

Broadcast Advertising Components
Internet Broadcasts / Video Blogs

Broadcasting on the internet is an extremely affordable method of advertising. Depending on skill level, agriculture marketers can often perform this type of promotion without the added expense of hiring a professional. With so many people having the capability to shoot digital video and then upload the footage to YouTube®, video blogs are becoming more popular. Many professional photographers and videographers include video posting on YouTube as part of their standard service.

Radio Advertising

This is one of the more overlooked advertising mediums, but one with tremendous potential benefit. Any operation planning to host an event with hopes of drawing a large local crowd (i.e. field days, annual bull sales) should definitely consider utilizing radio advertising. Spots can be purchased to air on a variety of stations from local broadcasts to syndicated farm shows across the region, state or nation. Radio advertising can be very affordable and targeted to a specific audience. For example, if marketing an annual bull sale, a producer might consider purchasing a 30 second radio ad to be played each morning during the farm report. This is a great way to reach local farmers and ranchers at an affordable price point.

Television Advertising

Television advertising is often one of the more expensive forms of marketing, but is becoming increasingly popular through RFD-TV. Producers have the option of purchasing entire episodes on popular broadcasts like American Rancher or Stock Show Confidential, or by purchasing commercial advertisements during popular agriculture related shows. Some associations like the National Cattlemen's Beef Association and American Angus Association host their own television shows with advertising opportunities available.

Web based Advertising Components
Blogs

Blogs, similar to an online diary, are a tool for individuals to document their daily activities online. Free software such as Wordpress® and Blogger® make it easy to create blogs featuring day to day activities, thoughts on life, and humor. The experience level required to use free blogging software is extremely low, and blogs can be easily updated from anywhere and everywhere...even your cell phones. Video blogging, which involves posting digital video of a blogger speaking, is also on the rise.

Email Blasts

Email blasts are an effective and affordable way of reaching large numbers of interested individuals, and in some ways replacing the need for print advertising. Email blasts distribute messages in a highly targeted, and fast delivery mode. Producers may utilize their own mailing list, or purchase the rights to a mailing list of an association or commercial marketing firm. Email blasts should be designed using a mixture of text and graphics (to avoid being considered spam) and include a very clear call to action like

V8 Ranch Email Blast

"Click here to for video and more information". By clicking the designated area, a link should then take viewers to a very direct website page, making it easy for clients to find the most pertinent information. Email blasts should also be shared on Facebook® pages and any other social media outlets.

It is very important to note that email blasts should only be sent to subscribers who voluntarily sign up to be placed on lists. Browsing through magazines or websites and collecting email addresses for blast purposes is illegal and unethical. If using a third party to distribute your blast, ask details about the size and demographics of their list circulation to ensure you are reaching the appropriate clientele.

Personalized Email Address

In conjunction with a company's website, personalized email addresses should be established (i.e. yourname@yourranch.com). This adds to the level of professionalism of your business and is one more way to keep the website address in front of clients.

QR Codes

QR code, which stands for quick response code, is a type of matrix bar code that was first developed for the automotive industry in Japan, cites Wikipedia. The QR code is an optical label that is read by a qr code reader, and directs the viewer to a designated web-baed location. For example, a rancher may place a QR code in his sale catalog that directs the reader to a YouTube link to the sale videos. The QR Code system has become popular in several industries outside of agriculture, however is still slow to catch on in the livestock sector.

Social media tools provide an excellent free resource to promote your business. These are also great ways to utilize the talents of ranch and farm children, who are often very comfortable using social media.

Social Media Promotions

There are many types of social media services livestock producers can utilize to continue to promote awareness of their product and brand; so many in fact that the time required to adequately utilize these services can be overwhelming. Most internet savvy livestock producers primarily use Facebook®, following with Twitter® as a secondary method of participation in social media.

In addition to the free components of social media, affordable advertising options are also offered through Facebook®. Facebook advertising allows users to select from highly targeted user accounts to get your message to those who are most interested in your product. For example, one could purchase a Facebook® advertisement that targets those users age 18 to 21 from Iowa who list showing livestock as one of their interests. Social media advertising is the fastest growing segment of livestock advertising.

Websites

For the most professional agriculturalists a website is the building block for all internet-based advertising. Websites provide a centralized presence for all online advertising including social media, video, blogs, email blasts and more. Every agriculture-based business should have its own URL name and website (i.e. www.yourranch.com). Websites are extremely affordable, easy to maintain and are considered a basic advertising necessity. A website

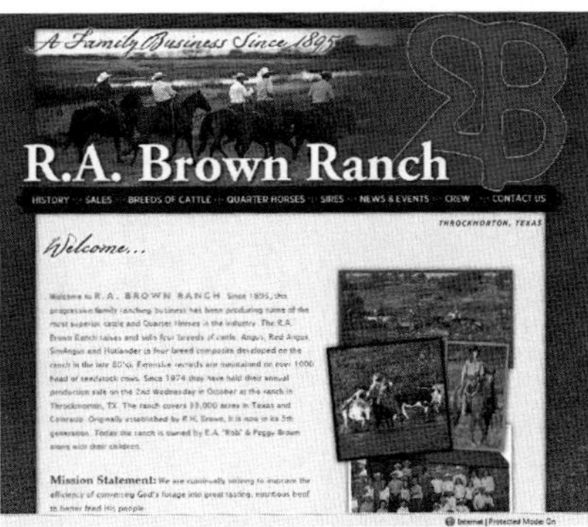

R.A. Brown Ranch website

should be one of the first components used in an advertising campaign, since ultimately many advertising calls to action direct users to a website for more information.

Website Banner Advertising

Some websites such as associations, news media, or blog sites sell web banner advertising as a way to generate income through their websites. Web banner ads are a smaller size, and their dimensions are based in pixel sizes to fit within the overall design of the existing website. Web banner ad pricing is usually told either by a specific time frame that your banner ad will appear, or by the number of impressions or clicks generated by your banner ad.

Since web banner ads are often very small compared to other types of advertising, keep the design simple. Your logo and website address may be sufficient. Or, use an animated flash graphic to create a moving web banner ad with faded photo transitions to help generate interest.

Ultimately, the goal of purchasing web banner ads is to drive traffic to your website. For this reason, it is imperative that your website be professional and updated at the time you purchase a banner ad. The goal is to send visitors to your web site, connect with them, then make a sale.

Web Links

Getting listed on websites other than your own add exposure to your business. Most livestock organizations offer a free website link to every member as an incentive of membership. If you aren't listed on your breed associations website, contact them for information on how to be added to their directory. Purchasing links on selected industry websites can also help increase the exposure of your website. Suggested examples of industry specific websites include:

Showpig.com is an industry specific directory that utilizes banner advertising to drive traffic to their clients websites.

- www.cattlepages.com - purebred beef cattle directory
- www.championdrive.com - sheep directory
- www.showpig.com - swine directory
- www.showsteers.com - club calf directory

Since the viewers of these directory sites overlap, we recommend getting listed on any free directories, and then choosing 1 or 2 paid directories for link placement.

Other Advertising Components
Apparel

As addressed in previous sections, establishment of company colors and a logo is an integral part of developing a marketing strategy. Once these two components are in place, it is advisable to consider purchasing customized apparel such as caps and T-shirts. For a more professional look, embroidered shirts or jackets are more expensive, but often worth the expense. Such apparel should be worn to all industry related events by owners and employees. This is just another way to keep a business name in front of a target audience.

Caps are popular tools for sire promotions, especially during key events like the National Western or other bull displays. We recommend Twisted Stitch embroidery or Showstring embroidery for your promotional cap needs.

Photography

Photographs are considered a very basic component of any livestock advertising campaign because without visual images it is very difficult to create interest and attention. Professional livestock photographers are readily available and affordable. Many people also enjoy photography as a hobby; so some livestock producers often have the necessary equipment and ability to shoot their own photos. The importance of good high quality photos cannot be emphasized enough, considering they are the basic component needed to draw interest and credibility to livestock advertising.

Promotional Give-A-Ways

Promotional give-a-ways can be used in various capacities to assist in name recognition. There are literally thousands of options for such material, but the more common ones include customized pens, note pads, calendars, koozies, caps and T-shirts. We recommend The Brand Company of Texas for promotional needs.

Tusa Show Cattle uses caps and T-shirts as promotional tools to help create brand awareness for their farm.

Sponsorships & Contributions

Charitable sponsorships can help create and maintain brand awareness. Though contributions are often made to organizations as a simple act of good will, they are also a form of advertisement. Any opportunity to be listed as a sponsor of events or awards will help create brand awareness and demonstrate to others one's civic contribution.

Text Messages / Picture Messages

Text messaging is a more modern tool for promotion, though they can be seen as distracting if over-used. Some associations, like the American Angus Association, offer text messaging services as a benefit of membership to keep Angus members informed of news and activities. Professional services, such as The Show Text, help broadcast messages to large numbers of people with a fast turnaround for a fee. These text messaging services allow for individuals to opt-in, and in some cases pay, to subscribe to industry related text messages. Texts are frequently sent out announcing show winners, judging contest winners, sale reminders, and sale reports.

Picture messaging tools, including services like SnapChat, can also be used for one-way communication to share pictures and videos with interested individuals.

So where does a livestock producer start?

Start with the building blocks: name, color, slogan, logo, Facebook page, business cards, and a membership in your breed association or professional organization. These tools lay the foundation for your marketing plan. Next, steadily add more tools to your toolbox as you are ready, starting with your website, email blast, and selective print advertising once your website is in place. From there...the sky is the limit!

SOURCES:

The Savvy Designer's Guide to Success, By Jeff Fisher – www.jfisherlogomotives.com, ISBN 1-58180-480-6. Published 2005 by HOW Design Books, Cincinnati, Ohio

The Guerrilla Marketing Handbook. By Jay Levinson & Seth Godin – www.gmarketing.com. ISBN 0-395-70013-2, Published 1992 by Houghton Mifflin Company

The Little Blue Book of Advertising, By Steve Lance & Jeff Woll. ISBN 1-59184-124-0, Published 2006 by Portfolio

The Fall of Advertising and the Rise of PR, By Al Ries and Laura Ries. ISBN 0-06-008198-8, Published 2006 by Harper Collins

Pantone Guide to Communicating with Color. By Leatrice Eiseman. ISBN 0-9666383-2-8. Published 2000 by Grafix Press, Ltd.

Sources:

Martha Garret, personal interview

Marketing Purebred Cattle, speech by Carlos X. Guerra

Why Advertise, by John Draper, Simmental Shield

CHAPTER 3:

Print Advertising

As our lifestyles continue to get busier and busier, our clients often ask if there is still value in print advertising? Today's busy lifestyle may make one believe that the future of modern agricultural merchandising lies solely on fast-paced social media or internet marketing. Do not be fooled, the power of traditional printed magazine or newspaper advertising can still never be underestimated.

When comparing the benefits of print vs. web advertising in the livestock industry, it is our recommendation that producers utilize both methods to maximize their exposure. We recommend utilizing digital advertising (website, social media) to provide constant, year-round news and updates for the business. We recommend selective print advertising during key times of the year such as at sale time or major livestock show seasons.

Print advertising offers advantages over web advertising in terms of long-term impact.

For many livestock producers, especially the older generations, there is no greater highlight than the first of each month when they receive their favorite publications delivered to their mailbox. Often these producers immediately drop everything to spend hours perusing and studying these printed publications, word for word.

Along those same lines, printed publications carry an extreme value of historical reference. Many farmers and ranchers save, if not hoard, each copy they have ever received of a favored printed publication. These back issues

LIVESTOCK PRINT VS. WEB

	NUMBER OF VIEWERS	CUSTOMER RELATIONSHIP	TARGETING ONLY SPECIFIC USERS	LONG TERM IMPACT	ANNUAL COSTS	DELIVERY SPEED
PRINT ADVERTISING	=	=		+	+	–
WEB ADVERTISING	=	=	+	–	–	+

make a handy reference for years to come and give advertised operations a lasting image that could quickly fade when relying solely on electronic advertising and promotion.

Printed publications also offer large circulations, with proven mailing lists of readers that are interested in specific products. This allows for targeted marketing to very distinct audiences. For example, those having a Hereford bull sale would naturally choose to advertise in the *Hereford World* as it ships directly to members of the American Hereford Association. Many producers also choose to advertise in the state or regional publications of their target audience. By selecting the printed publication most closely aligned with the target clientele, a marketer can specifically reach thousands of readers who are directly interested in what they have to offer.

Where do you advertise? There are hundreds of printed media outlets that focus on the livestock industry, and purchasing print ads in multiple publications can quickly take a large chunk out of an advertising budget. Printed publications can be broken into these broad categories:

- Local newspapers – *Wharton Journal Spectator, North Platte Telegraph*
- State breed association publications – *Texas Hereford, Missouri Angus*
- Regional publications – *Ozark Farm Neighbor, Western Livestock Journal*
- Segment specific magazines – *The Showtimes, The Showbox*
- National breed publications – *Angus Journal, Shorthorn Country*
- Beef industry publications – *Beef, Working Ranch*

Case Study: The Exposure Sale, by Christy Collins

The Exposure sale, which grossed over $1 million, is an example of an ideal purebred livestock sale advertising campaign. This particular ad, "More Champions. More Dollars Generated" utilizes every strategic component of good advertising. First, the ad is a two-page, color advertisement, which makes it highly noticeable in publications of any size. Collins uses a headline, photos, ad copy, a call to action, and a signature to capture the readers attention and direct them through the advertisement.

The ad is extremely well designed. The large dominant photo across the top instantly captures your eye. The five photos in the center of the page show examples of the sale offering. The ad copy describes the benefit of purchasing cattle from The Exposure sale (high sellers and champions). The ad concludes with a call to action of "Request a Sale Book Today".

Before placing a print advertisements, livestock producers should first have a professional, updated website in place. Print advertisements are constrained to specific page dimensions, which often limit the number and size of photographs and text that can be used. A website, on the other hand, can contain unlimited photos and text. All printed advertisements should direct readers to a website where they can obtain a much greater amount of information on the business.

Every livestock producer should run printed advertisements in their target publications, even if the advertisement is small. Printed advertisements serve as a reference for others who may be looking for your contact information specifically, or looking for breeders in their area. Every purebred livestock producer should run, at minimum, a business card or state co-op ad in their respective breed association publication or state cattlemen's association publication. These ads run at a minimum cost and give your operation a quick reference to your fellow producers.

Four Ways to Get an Ad Noticed

The cost of print advertising can quickly add up, therefore it is important to make every dollar count by maximizing design strategies. The following four suggestions are a good place to start:

1. Use Color

In a publication that uses mixed ink, color print stands out over black and white any day. The use of color increases an advertiser's chance of getting noticed in a publication, since the human eye is naturally drawn to it. If a four-color ad is too expensive for your advertising budget, or you are advertising in newspapers, ask the publication about the option of using one-color (spot color) to give the ad a little pizzazz.

Color advertising becomes more important when ads are placed in publications with 100 pages or more. As the size of a publication increases, advertising retention decreases, simply because the reader is faced with more advertisements competing for their attention. In publications with 100 pages or more, a black and white ad, or even a half-page ad will most likely be passed over by the consumer; so color and at least a full-page ad are a must!

When advertising in publications with less than 100 pages, or newspapers, color ink becomes less important, and overall readability of the ad takes precedence over color ink. State cattlemen's association publications are a good exception to the rule of thumb on the importance of color vs. black and white advertising. Many state cattlemen's magazines are primarily black and white publications. In these publications, color can be foregone to save advertising dollars as long as the ad is very clear and easy to read.

2. The Bigger the Better

Whenever possible, purchase the biggest size advertisement you can afford. If advertising in publications with 100 or more pages, we recommend purchasing a full page advertisement for maximum exposure. A two-page ad is even more advantageous, especially in publications with more than 200 pages. A two-page spread allows for more photos, larger photos, and more space for easy-to-read captions and headlines. The side-by-side spread of a two-page ad also greatly increases the chances of being noticed and remembered by readers since the ad occupies an entire page spread, and therefore does not compete with another ad for attention.

On very special occasions, or in publications with more than 300 pages, inserts may be used to draw an even larger amount of attention to an advertisement. Consider purchasing a foldout, tear-out catalog, ride-along catalog or having an advertisement printed on heavier weight paper to capture reader interest.

While these recommendations can greatly help improve notice ability, an ad of any size — even very small black and white ads — can be effective if is has the correct information and a strategic design that drives readers to the business website to obtain more information.

3. Use Photographs

Displaying photos of a product gives people a reason to stop and look at an ad, and more incentive to pick up the phone and call, or better yet get in the car to see the product in person. Photos are a component in advertising that can make, or break any good ad.

It is imperative to remember that any photo used in printed advertising must be of high resolution, preferably able to be scaled to 300 dpi and a size of at least 5x7 inches; the bigger the better. Never save a photo from a website and attempt to use it in a print ad.

Some of the more popular web-based email services will automatically downsize photo attachments in efforts to make the emails load and send faster. This causes problems for magazines and designers who need the original file size photo. Thus, it may be necessary to adjust email attachment settings to ensure an outgoing photo will remain at the original size.

When a budget will allow, hiring a professional photographer, who specializes in the advertised product, will ensure that photos are high quality and can be used in print or web mediums. Photographers are readily available, with prices starting from a few hundred dollars to thousands of dollars, depending on their experience level. Always schedule photographers far enough in advance to allow enough time to shoot the photos, edit the photos, and send them to the appropriate designers.

4. Use Testimonials and Examples

Tell, and show your reader why they should be using your product. For example, if you are holding a showpig sale, use photos of recent show winners in your advertisement to show a record of success. If you are running an ad to promote a sire, show examples of progeny.

Lautner Farms, a club calf operation in Iowa, used this technique, under the direction of Ranch House Designs, Inc.©, in the late 2000s to showcase various breeder success using Lautner Farms bulls. They built an entire advertising campaign using the theme of "Lautner Bulls Make 'Em Easy to Sell...and Tough to Beat.'" This campaign was widely featured throughout show publications, and included lengthy lists of different clients with the prices their Lautner-sired calves had demanded. The ads also included numerous photos of show winners sired by Lautner bulls. The farm found that many of their customers enjoyed being featured in the advertisements, and looked forward each month to seeing their names in print.

The Five Things That Should Be In Every Advertisement

As with any type of printed advertising, the focus of an ad should be the product. However, while browsing through any one of today's popular printed media publications, one can become overwhelmed with the number of advertisements competing for consumer attention. Some publications feature extremely trendy, often overly-busy ad designs that can leave the reader with a headache. Other designs may be so boring and dull that they put the readers to sleep.

This ad for Dismukes Ranch features all of the required components for an outstanding livestock advertisement: a headline (Real World Cattle for Real World Cattlemen), a photo, two paragraphs for ad copy, a call to action (welcome you to visit the ranch), and contact information. This ad was placed in the Charolais Journal calendar thus the sideways orientation. We do not recommend sideways ads in regular format publications other than calendars.

Throughout history, the basics of good advertising, agricultural and otherwise, have revolved around these five components:

- A headline
- A photograph or visual element
- Advertising copy (text)
- A call to action
- Contact information

These elements interact together to create the perfect advertisement, and help achieve the desired result. They assist the reader in understanding the advertising message and upon viewing leave them with a feeling of fulfillment.

The headline should be creative, concise, and most importantly, clearly legible. A good rule of thumb is the headline should be at least three times larger than the other text on the page. The headline sets the overall tone of an advertisement and prepares the reader for the intended message of the ad.

Headlines may be anywhere from one to ten words, most of which should be creative and attention-grabbing. For sire promotion, quite often the headline of an ad is simply the animal's name; in sale advertisements the headline may be the name of the sale. If themed graphical elements are used in the design, the headline should closely integrate with the photo to tie the graphic element to the message.

Sample headlines include:
Success Starts with SEK Genetics
Legends are Made at Bright-Leo Show Cattle
The Dismukes Advantage

Photographs are imperative in agricultural advertising design. Preferably, the photos used should be of the product being marketed. However, in the undesirable event that no photos of the actual product are available, some type of stock photo or background should be used to include a graphical element.

Livestock photographs in advertising should be as large as possible, clear, and unobstructed from view. Remember that the goal of the advertisement is to show off the product to potential clients. Each photo should include a concise and easy-to-read caption explaining the relevance of the photo to the ad. If picturing a specific animal, include the animal's name, pedigree and other necessary information.

Every good advertisement will include some type of advertising copy, or text, explaining the purpose of the ad to the reader. This text can range from a single sentence to several paragraphs, depending on the advertisement's purpose. Typical text in livestock ads includes highlights, such as an animal's pedigree, performance information, show accomplishments, notable siblings and other selling points.

Regardless of your design preference, clarity and legibility of print advertising is very important. All fonts and type styles on the ad should be easy to read. If possible, keep all of fonts no smaller than 10 point in size, enabling audiences of all ages to read the message.

Designers can also use the 3:2:1 ratio when determining font sizes within their advertisement. This ratio explains the size relationships between the headline, subheadings, and ad copy. Using this ratio, a headline may be sized at 48 point font, subheadings at 24 point font, and ad copy at 12 point font. This helps direct the reader to recognize the areas of importance in the ad and creates a natural flow from headline to subheading to ad copy.

When writing advertising copy, the options are to open with a statement, open with an invitation, or open with a question. For example:

Opening with a statement

For over 25 years Duelms Prevailing Genetics has been a leading source for consistent genetics that yield championship results! In the past 12 years we have produced 28 Grand or Reserve Grand Champion Barrows at Texas majors! We have pigs available all year around for most county shows, and majors in the country. We pride ourselves on customer service and satisfaction, and will do whatever it takes to earn and keep your business. We have over ? years of barrow feeding experience and will help you in whatever way we can during the feeding period. In addition to our showpigs, we offer semen on our leading barrow-producing sires. Our boar stud may be small compared to some, but unlike most, we utilize all of our boars in our own genetic program. What does this mean to you as a customer? Consistency, Predictability, Results. Pigs for sale all day, every day.

Opening with an invitation

Elliott Cattle invites our southern friends to view our online offering of the best bred heifers we have ever offered! Most are out of Irish Whiskey, Chill Factor or Simmental dams, and AI bred to Ali, Jesse James and Carpe Diem to start calving February 2012. Here is an opportunity to buy top Midwest genetics from a proven home of champions! We are also happy to assist in boarding arrangements of purchases by those in drought-stricken areas.

Opening with a question

Do you need a better bull? Dismukes Ranch Charolais bulls are performance tested as yearlings, then placed on a hay-only diet. We focus first on feet, fertility and dam's udder quality; then, on visual quality and actual performance. Lastly, we focus on EPDs. Without the first two, it doesn't matter what EPDs are! Our Angus EPDs are at or above breed average for all

16 of the commonly used EPDs. The only exception is our $EN, which is 3X the breed average. Our bulls are moderate framed, stout, deep bodied bulls that will add pounds in an optimum frame size built for longevity. All bulls' semen tested, negative for PI BVD and anaplasmosis vaccinated. Free delivery in Oklahoma and surrounding states. Sight unseen purchase guarantee. Repeat buyer discount up to 10% off.

The call to action is an often overlooked, but a necessary ad element. It gives the target reader the take away message of the advertisement. The call to action gives the reader an idea of the next step to be taken, after reading the ad.

Example calls to action include:
Call Us Today to Place Your Semen Order
Please Join Us September 1st for Our Annual Sale
Visitors Welcome Anytime, Call Today to Schedule a Visit

Finally, every ad should include contact information. Depending on the advertiser's preferred methods of contact, this can include anything from a name and home telephone number, to an email address, website address, fax number and even cell phone numbers. Remember to make it easy for prospective customers to reach a knowledgeable person. If email address is rarely checked and responded to, be cautious in listing it in advertising. When clients use email as a method of contact, they typically expect a response within one business day.

After these five required elements have been addressed, a marketer can then choose the style of advertising he or she prefers. Some prefer very intricate ad designs, while others prefer a more minimalist approach to design. It is important to consider the advertising preferences of the target audience. If an advertisement is promoting a purebred commercial bull sale in western Nebraska, chances are that particular audience would not appreciate an overly-trendy design with a grunge look. Conversely, if the target audience is youth livestock exhibitors, they would be turned off by tiny pictures and an entire page of performance data captions under each photo.

A "cool" looking ad may catch a reader's attention but doesn't necessarily equal more sales. Remember to use large, clear photos combined with captions that are easy to read. Also, insure contact information is clearly legible and easy to locate.....after all, the goal of your ad is to get potential clients to call you!

Ogilvy vs. Z Layout of Design

There are two general advertising layout styles commonly found in agricultural advertising: the Ogilvy format and the Z-format.

The Ogilvy format was named after advertising expert David Ogilvy, who used this simple block format as the basis of his most successful advertising designs. Using this setup, a designer can create ads in a simple top to bottom block format within the page. The headline appears at the top of the page, followed by a large photo which is followed by text that contains the call to action, and finally the signature.

The Ogilvy format is extremely popular in sire and stud advertising,

"4MF Movin' Up" is an example of an Ogilvy format ad. It utilize one large dominant photo that takes up approximately half of the page. The headline is at the top and the signature along the bottom of the page.

which usually includes a photograph of the animal that claims up to half of the page space. This format can also be altered and modified slightly with the photograph appearing at the top of the page, followed by the headline or other similar modifications. However, it is easy to recognize an Ogilvy format design due to its simplicity, clarity and large dominating photo or graphical element. This format is also best suited for advertisements that use one large photo and possibly a small supporting photo.

The Z-layout is a modern design strategy based on the principle that a reader begins looking at a page in the top left hand corner, moves their eye across to the top right, down the page in a diagonal line, then exits the page on the bottom right hand corner - in a Z format.

Z format is a great design style to use for advertisements that contain multiple photos, allowing the photos to be arranged strategically down the page, creating a nice flow. With this configuration, the headline still typically appears at the top of the page, photos and captions move down the diagonal line of the Z, with supporting text appearing in the bottom left corner and the signature and contact information in the bottom right hand corner. The strategy behind this positioning is to leave the reader with the company logo and contact information, as well as making it easy to find.

Designers using either of these formats should pay close attention to contrast, repetition, alignment and proximity (CRAP) as outlined by Robin Williams in The Non-Designers Design Book (3rd edition, 2008). Contrast deals with the visual attraction on a page, and is usually what first catches a reader's attention. Contrast can be achieved by using contrasting colors and fonts along with large visual images to naturally direct the reader where their eyes should first look.

"Classic" for J.D. Hudgins, Inc., utilizes the Z layout. This is recognized by the headline in the top left corner, which guides you to the leather tag in the upper right corner. The eye is drawn diagonally down the page for the 3 photographs, and ends with a signature that fills the bottom portion of the page.

Repetition is the process of using repeated visual elements throughout the design, and also to use the same general look for the company in all advertisements. For example, use the same fonts and colors in your advertising each month to build a corporate identity. For example, each year, V8 Ranch selects 3 fonts that they will use for the entire year in all of their materials: an accent font, a headline font, and a copy font. No matter what project is being created, these fonts are used. The ranch also has three colors that are used on everything: navy blue, white, and gold for an accent. They strictly adhere to these graphic standards so that readers instantly recognize their advertising by these repeated elements.

Repetition of fonts, colors, a large photo, and consistent design helps build the recognizability of V8 Ranch advertising.

Repetition can also be used in overall design strategies. V8 Ranch places a 1/3 page black and white ad monthly in the *Gulf Coast Cattleman.* This smaller ad in a black and white format offers an affordable way to keep their name in front of the commercial cattleman. They use a different ad each month, but the general format of the ad remains the same: a large photo, the same font, and the same general amount of text. As they repeatedly run their ads, people begin to recognize the brand image.

Alignment deals with the way that elements, text and objects are strategically aligned to create a clean look. There are four types of generally accepted alignment styles: left, center, right, and justified.

Right alignment puts all of your text in line with the right margin.

Left alignment puts all of your text in line with the right margin.

Center alignment centers all of your text between the left and right margin. Center alignment should be used only for captions or invitations.

Justified alignment makes all of your text in a block format. Used correctly, it can be very professional, but it also can cause problems with letter spacing as words are spaced out if full justification is used.

Finally, proximity helps group related items together to help organize the information and make it easy for the reader to see a clear structure. Proximity is important when choosing the placement of pictures in a layout.

Typography

Readability should be a major priority of good advertising design, second only to the focus on the actual product. Fonts should be clearly legible and large enough to be easily read by people of all ages. If someone can't read the company name, a sale date, or the contact phone number, what is the point? There are thousands and thousands of fonts available. The following classifications are the most common:

Serif Fonts	Times New Roman
	Garamond
Slab Serif Fonts	**Clarendon**
	Century Schoolbook
Sans Serif Fonts	Arial
	Myriad Pro
Script Fonts	*Brush Script*
	Chopin
Decorative / Grunge Fonts	**Bleeding Cowboys**
	Dear Joe

The number one typography rule to consider when designing: Never use Comic Sans.

The second design rule of thumb concerning typography: Never use more than three different typefaces in a single design. Traditionally, a designer will use one font for the headline or accent font. This font is used in the headline, subheading and any design elements that call for special attention. The second font is your copy font, which is used for ad copy, captions and contact information. Typically the copy font is a basic, easy-to-read Serif or San Serif font such as Garamond or Myriad Pro.

Other basic guidelines for typography:

- Never use two different script fonts on the same page.
- Use variation in typefaces to create interest. To achieve this, use bold text attributes, italic or variations of font size. However, don't bold EVERYTHING on the page. Use these attributes selectively to create

contrast and direct the reader.

- When using a fancy, grunge or decorative font on a design, choose a second font that is clean and easy to read, like Serif or San Serif font.
- Do not center all of the fonts on the page.

Different types of advertisements

Once the decision has been made to place a printed advertisement, often the challenge lies in deciding exactly what will be featured in the allotted space. Following are some common types of advertisements used in agricultural marketing and the components they involve.

Sire ads are used to promote semen sales or herd sires. They typically feature one large photo of the sire, along with an eye-catching graphic that coincides with the animal's name. The graphic helps to reinforce the name and identity of the animal and creates a memorable look. At minimum, sire ads should include the name, registration number (if applicable), sire, dam, birth date, performance information, semen price and contact information for placing an order. As the sire is more widely used, progeny photos can add value to the advertisement to show the breeding merit of the

"Royal Stockman" for Topline Farms and Vision Angus, sire ad.

"The Little Team That Could" for Teague Family Shorthorns, Congratulatory ad.

sire. Photos of full siblings, sire, and dam are also useful considerations to include.

Winner / Congratulatory ads showcase competitive winners raised by or owned by a farm or ranch. These type ads are also useful in creating name and face recognition as a tool for helping show judges notice certain showman or certain animals. These are more common among purebred breeders due to abundance of competitive shows and usually appear in breed publications or youth publications. Many producers marketing livestock to junior exhibitors use this strategy to educate potential customers on the show ring success that can be purchased from their particular operation. Most people enjoy seeing their own photograph in print; making these ads extremely useful as a public relations tool by recognizing a business's valuable customers.

Thank You ads are another way to show appreciation to your buyers. These ads effectively accomplish two goals: 1) They make your buyers feel important and 2) They show your fellow producers who buys your livestock. This helps make a seller appear approachable, easy to work with, and encourages others to buy from you.

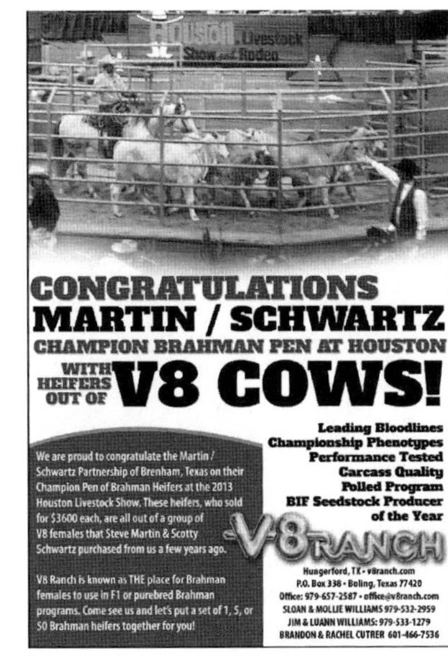

This thank you ad congratulates and thanks a repeat buyer from V8 Ranch.

"J3 Cattle Company" used this ad as a combination winner / sale ad.

Reputation building ads are used to keep reminding readers and fellow producers about the core mission and values of a business. For new operations, this type of ad can be used as an introduction to the particular industry and make people aware of important philosophies, goals, etc. For established operations, these ads can reinforce what the company is known for. Reputation building ads are best used in publications that serve as a reference or that are used year-round, such as a state breed association membership directory. Reputation ads are very general in nature, and thus never become outdated. They often include scenic photos, general pasture pictures, basic philosophy and contact information.

Sale ads are used for the purpose of making potential clients aware of an upcoming sale. They should always include a very prominent feature of the sale name, date, time and location. Good sale ads include

Circle A Ranch uses a reputation ad to build their brand image and encourage readers to attend their sale.

Vision Angus' sale ads prominently feature the sale date, time and location along with a great preview of the sale offering and performance data on featured lots. It also lists AI sires featured in the sale to give buyers an idea of the quality offered.

a short description of the offering, such as number of head or lots offered, breeds of animals, and classes of livestock selling (i.e. bred females, breeding age bulls). It is very important to include photographs of the sale offering to help generate interest. If photos of the actual offering are not available, use photos of high sellers from past sales, or reference animals such as service sires to help attract interest. Be sure to include a contact for catalog requests and a call to action, inviting readers to the event.

Extreme design ads are those 'off the wall' type ads that include a very abstract design and generally do not follow the general standards described in this book. When used

"Hitting the Mark" for Bulls Eye Ranch is a minimal design that utilizes a bold, attention getting tactic of extreme design.

sparingly, they can make a big impact with their shock value and do have some value in getting an ad noticed. Extreme ads often focus on a very dominant and bold non-traditional graphic or headline. Or, they may involve something as shocking as running an ad upside down in a magazine. Keep in mind that while they are effective for one time use, or very sparingly, a successful advertising campaign cannot be built around extreme design ads alone.

Occasionally you will see advertisements in print publications where the ad is turned sideways, or even placed upside down, as an attempt to be different. These strategies are usually fruitless and should be avoided. Quite simply, many readers are too lazy to actually turn the magazine around to view the ad, and thus your ad never gets read.

Extreme ad designs can be fun to design, and give a fleeting shock value that creates hype, but do not provide the long-lasting impact that a well designed ad with photos and informative ad copy provide.

Minimal design ads are ads that are built around the principle of less is more. These ads can be very visually appealing but should be used sparingly,

"The Bold Ribbed & The Beautiful" by Christy Collins for The Exposure Sale. This sale advertisement was a 2 page spread. This distinctive design is an example of a very effective use of a minimally designed ad.

as often times they do not follow the basic design principles of using a headline, copy, photograph, call to action and signature. A minimal ad may be one that simply has a single photo with the company logo and website URL. Or it may be a 2 page spread that simply features one bull photo. White space is utilized to create a very aesthetically appealing ad with very little text.

Planning ahead for advertisements

While some people work well under pressure, it is always good to plan in advance to allow a publication enough time to create an outstanding advertisement. This is even more crucial during busy times of year like sale and show seasons. Masterpieces are rarely constructed two days prior to a print deadline. Advance planning is even more crucial when shooting photos; rainstorms and snowstorms have a funny way of sneaking up at the last minute and delaying production time. It is never too early to begin the planning process, contact designers, book photographers or reserve magazine space.

Ideally, a business should create an advertising plan for each year. A typical yearly advertising plan might include a monthly ad in a breed publication, combined with a heavy advertising push during a production sale season in other publications. It is a good idea to give yourself and your design anywhere from 4 to 6 weeks to complete a printed project. Creating a plan assists in determining when advertisements are due to certain publications, and helps ensure photos are taken and materials are ready by each deadline.

SOURCES:

The Non-Designers Design Book, Third Edition
By Robin Williams
ISBN 13: 978-0-321-53404-0
Published 2008 by Peach Pit

Words that Sell, By Richard Bayan
ISBN 0-8092-4799-2
Published 1984 by Contemporary Books

Graphic Design Cookbook, By Leonard Koren & R. Wippo Meckler
ISBN 0-8118-3180-9
Published 1989 by Keonard Koren & R. Wippo Meckler

Merchandising Bulls at Private Treaty, by Martha A. Hollida. Brangus Journal,
    September 1982.

CHAPTER 4:

Brochures, Cards, Catalogs, & Direct Mail

Brochures, direct mail and catalogs are the next rung in the printed advertising material ladder of progression. Agricultural marketers generally utilize these tools once an effective print advertising campaign is underway. This print medium most often acts as a stand-alone piece distributed at special events, the company headquarters or through the postal service.

The advantage of a catalog or brochure is the absence of competition in the material; as opposed to a magazine advertisement where numerous other advertisers are competing for readers' attention. These individual pieces isolate and draw attention to a business's specific product and brand, while providing potential clients with exclusive information.

Direct Mail

A tri-fold brochure is the simplest of all stand-alone direct mail pieces. Traditionally these are printed on 8.5x11 inch paper, and folded into three sections. This configuration is small enough to mail, fits easily in a shirt or back pocket, yet large enough to provide two pages of information in a creative and professional package.

The standard layout for a tri-fold brochure is a basic six panel design. Conventionally, the front panel contains the business logo and an artistic element of a photograph or simple background. In the same respect, business contact information lends itself nicely to the back panel. Equivalent to a full page at landscape orientation, the inside three panels are generally where the bulk of information and photos appear. The remaining inside flap panel is also available for brief text and small photos.

Inside Front Cover   Back Cover   Cover

Inside 3-page spread

Wildcat Creek Ranch, by Christy Collins, utilized a 6-page brochure. This design was achieve by a tri-fold on a 25 inch wide by 11 inch tall piece of paper, giving you 6 panels for information.

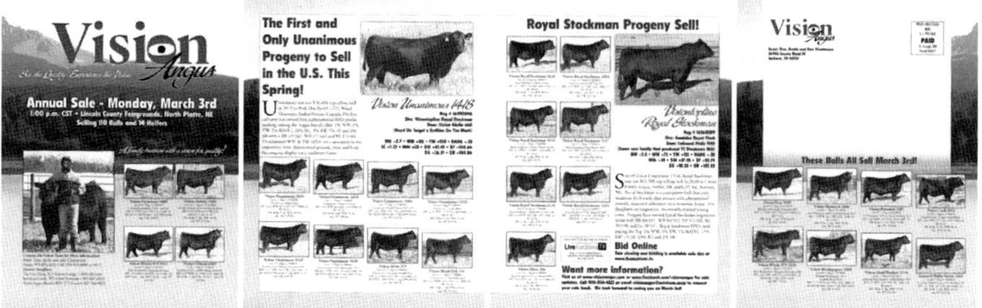

Vision Angus utilized a 4-page brochure direct mailer to promote their 2014 bull sale. The cover featured the sale information. The inside spread featured two of their leading herd sires and his progeny selling. The back cover included a mailing area and 6 pictures.

Aside from the standard 8.5x11 tri-fold format, brochures can take on all shapes, sizes and folding options. Legal size paper of 8.5x14 inches can be used for a four panel design. The variety of brochures available is endless, including the accordion fold, gate fold, single fold and many more. Of course, the more elaborate the folding and odd the paper size, the greater the printing expense will be.

Postcards make a great direct mail marketing item at an affordable price. Standard postcards are 4x6 inches and qualify for the US Postal Service discounted postcard rate. This is a popular option for those wanting to do a quick mail piece, without incurring a huge printing and postage expense. An oversized postcard is 5x7 inches in size, and although it makes more of an impression, it also requires first-class postage. Postcards are very popular for sale reminders, announcements and sire promotions.

Dismukes Ranch utilize direct mail postcards as an affordable promotional tool for their annual bull sale. These postcards are designed to be self mailers.

When intending to mail cards, the entire front side is available for information; but only one-quarter of the opposite side is available, and even that portion must include the return address. The remaining portion of the back panel must remain blank to allow for a mailing label, postal bar code and stamp.

If the intended use of a card is for handout purposes only, or if it will be placed in an envelope, more room is available for promotional information. This type of postcard is popular in sire or stud promotions, often seen at show events and field days. Usually the front side of the postcard features a single large 5x7 inch photograph of the featured animal, and the back side of the card includes the pedigree, a paragraph of information about the animal, and contact information of the owners.

Catalogs

Catalogs are possibly the largest, most expensive and time consuming advertising project an livestock producer will complete in the course of an advertising program.

Organizing

There is no such thing as starting too early when it comes to planning a sire catalog, donor catalog or any type of sale catalog. The recommendation is to begin the process approximately two months before the catalogs are expected in the hands of clientele. Using the indicated sale date, the following time line is a good example:

Sale Date:	**March 15**
January 15	Begin organizing materials, taking pictures and writing footnotes. Allow two weeks for this process.
January 28	Completed sale information available to catalog designer and sale manager. Begin assembling catalog mailing list.
February 3	First proof received from designer. Necessary changes reported back to designer or manager.
February 10	Second proof of catalog received. Necessary changes reported.
February 15	Final proof received and approval given for print. (Approximately 1 month prior to sale)

February 18   Catalog sent to printer and/or publication with mailing list.
                     Catalog immediately published on business website.
                     Email blast is sent announcing catalog availability online.

February 25   Catalog printing complete and copies mailed.

March 1   Catalog arrives in client mailboxes.

March 10   Email blast reminder sent

March 15   Sale Day!

A successful catalog design project includes the following information:

- List of sale lots
- Photographs of individual lots (usually in high resolution .jpg format)
- Supporting photographs if necessary (sire, dam, progeny, etc.)
- Pedigrees, performance information (usually typed into Microsoft Excel® or retrieved from a breed association in a .csv or .txt data file)
- Breeding information (AI date, exposed date, etc.)
- Footnotes (4-6 sentences describing the lot, usually typed into Microsoft Word®)

The list of sale lots, or lotting list, should be finalized, and in most cases set in stone, before beginning the catalog design process. While lot assignments can change and offerings added or removed after the catalog has been designed, it is not easy and designers generally charge a fee for excessive changes. If at all possible, finalize and maintain the lot assignments to avoid confusion among all involved.

Lot Information

Ideally, each sale lot should have a photograph accompanying the lot information. When auctioning livestock, the photo should be of the animal being offered. Support photos such as sire, dam, full siblings of merit and progeny of merit may also be included. Photos should be of high enough resolution to scale to 300 dpi and are typically saved as JPG images. Photographs used in color printing should be formatted as CMYK (cyan, magenta, yellow, black) photographs and those used in a black and white catalog should be converted to grayscale.

Include a pedigree for each animal in the sale. Catalog designers will eventually need this information in an electronic format in order to input the data into the catalog design software. Such data can usually be obtained in an Excel® spreadsheet or CSV (comma separated value) text file from a breed association or pedigree service. While there may be an additional fee for obtaining this electronic file from the provider, it is most definitely worth it. Obtaining data from a reputable service ensures it is free from typographical errors and includes the most updated information available. It also saves tremendous amounts of time by not having to locate each pedigree and re-type it into an Excel® spreadsheet.

Most catalogs include a two generation or three generation pedigree of each animal. However, purebred breeders place heavy emphasis on pedigree merit, therefore it is more common to see a three generation pedigree in purebred sale offerings. When non-pedigreed animals are offered, the sire and dam may be listed.

Breeding information on female lots should be specified when applicable. Such information would include the AI date, or exposure date if naturally serviced, and the service sire. If a female has been determined safe-in-calf, include that information as well.

Performance data is a necessary component of purebred livestock sale catalogs. Depending on the species and breed, this may range from a few data variables up to thirty data fields. Purebred sale catalogs typically list:
- Animal name
- Registration number
- Date of birth
- Sex
- Tattoo or identification number
- Horned, polled or scurred status
- Hide color
- Actual weights (birth, weaning, yearling)
- EPD values for BW, WW, YW and Milk

If a particular association or breed offers additional EPD measurements, list those as well. Sire catalogs may also include ultrasound data, ratios, scrotal circumferences, calving ease predictions and adjusted weights. Information is power; if the data is available always include it.

Writing Footnotes

Finally, each lot in a sale catalog should include a footnote. This may be as simple as a few bullet points outlining strengths, to multiple paragraphs explaining in detail an animal's pedigree, progeny, siblings, etc. A good place to start would be the following format:

- One opening sentence to catch the readers' interest and summarize why they should buy the animal
- One or two sentences describing the sire bloodline and performance
- One or two sentences describing the dam bloodline and performance
- One or two sentences describing the animals physical conformation
- One sentence describing performance information if noteworthy
- Optional sentence(s) describing special performance of any full siblings or close relatives
- Optional sentence(s) listing show ring performance of the offered animal
- Optional sentence listing any special sale terms (half interest, retaining a flush, guarantees, breeding status, etc.)

"Maternal Legends 2011" by Arin Strasburg. As seen in this example, each animal in the sale is represented by a photograph, animal information, and footnotes. Once this general layout is established, it is repeated over and over again for each animal in the sale. Much of this typesetting can be achieved through using the data merge feature in InDesign®.

A typical footnote for an elite purebred sale would read like this:

Look no further for a gorgeous bred cow with lots of breed character, destined to be a great producer. She is a daughter of the now deceased +JDH Sir Avery Manso 159/6 who is known for siring outstanding females with tremendous length of body and maternal characteristics. Her dam is the Register of Renown +Miss V8 933/5, who is the epitome of V8 style and has been featured in many ads through the years. 933 is a Powerstroke daughter out of the Dickens' +Miss V8 4/5 cow (a Suva daughter!) Lots of great Brahman names in this pedigree! This bred cow is also a maternal sibling to Miss V8 416/5 ($50,000 Houston high-seller), Mr. V8 111/6 (herd sire for Fat Dog Ranch) and Miss V8 23/7 (many time champion 2008-2011). The only reason she sells is because we are keeping her flush mate, Miss V8 23/7. This cow comes from genetics that are the backbone of our program and will no doubt add value, production and name recognition to her new owner. She sells bred to Mr. V8 380/6. The Avery x 380 crosses have been exceeding every expectation we have ever dreamed.

Footnotes may be considerably shorter in large-volume bull sales, like this footnote taken from the 2014 Circle A Angus bull sale catalog:

Third heaviest WW EPD, heaviest YW EPD and highest Terminal Profit of CAPB, WW Ratio 103, YW 111, IMF 116, REA 103

Bulletted footnotes are also popular for concise attributes of the animal selling, like this one utilized by Sullivan Farms:

- *A real stunner!*
- *She will be a load to get around in the show ring and will then turn into a great investment in the pasture for generations to come.*
- *It is hard to argue with the consistency of the Myrtle Bo 46P progeny.*
- *A maternal sister to the 2011 NWSS Champion Female for Hannah Moore.*
- *Registered purebred Shorthorn/Registered MaineTainer*

In many non-pedigreed sales, footnotes are not as lengthy due to the lack of detailed information and performance data. In this case, it is perfectly acceptable to keep footnotes brief or arrange them using simple bullet points.

Layout

Once the content information is assembled, the actual design and layout of the sale catalog begins. The lot layout will vary depending on the complexity and size of the catalog. One lot per page is considered a premium layout, but often a page will include three or four lots per page when photographs are used, and six or eight lots per page without photos.

Smaller sale catalogs often use a basic template layout, based on a page divided into six or eight boxes, allowing for pictures and text of three or four lots per page. If photographs are being utilized, they should be placed to the outside of the page. This allows readers to view them with ease when flipping through the catalog.

Catalog design styles must facilitate easy reading and legibility. Refrain from solid black backgrounds with white type. Avoid grunge or distracting backgrounds that prevent readers from clearly reading the text. The overall goal is showcasing the animal and providing clear information to potential buyers.

Top Line Farm 2013 Sale Catalog Cover

Top Line Farm 2013 Sale Catalog Information page

Covers and Additional Pages

Most all catalogs include a few basic, yet necessary pages in the catalog design, such as covers, welcome letters, sale terms, and sale information pages.

The sale name, date, time and location should be clearly displayed on the front cover and back cover, enabling viewers to find the pertinent information quickly and easily. Regardless of which way the catalog falls on a counter or table all important points will be displayed.

Customarily, the first interior pages of most catalogs include contact information, sale day telephone number, associated staff, official auctioneer, the sale terms, a welcome letter, local hotels and travel information, schedule of sale events, livestock representatives and any other general information.

Many breed associations offer a standard sale terms and conditions document that can be used to cover all terms, conditions and guarantees that should be included in the catalog.

The back cover of a sale catalogs intended to self-mail must be designed to fit postal regulations. Currently, postal regulations stipulate that the top one-third of the page be solid white to allow for bar code scanning and automated mail service. Check with your printer or designer to ensure your catalog complies with these regulations.

Printing

When determining the cost of printing a catalog, five factors come into consideration:
- Number of pages
- Paper weight and style
- Color of ink (black and white vs. color)
- Quantity to be printed
- Binding options

By definition, catalogs are individually printed materials with four or more pages that are bound together. Due to the nature of printing processes, all catalogs must be designed in four page increments. With that said, a catalog's page number will always be 4, 8, 12, 16, 20, 24, 48, 60 and so on. This page count also includes the cover pages.

Paper weight and style is another price determining factor. Standard catalogs are printed on eighty pound (80#) gloss or matte paper. This is the common paper style and weight seen in most magazines. The term "self-cover" is used to describe catalogs covered using the same paper as the interior pages. Heavier weight paper can be used on the cover pages to increase quality.

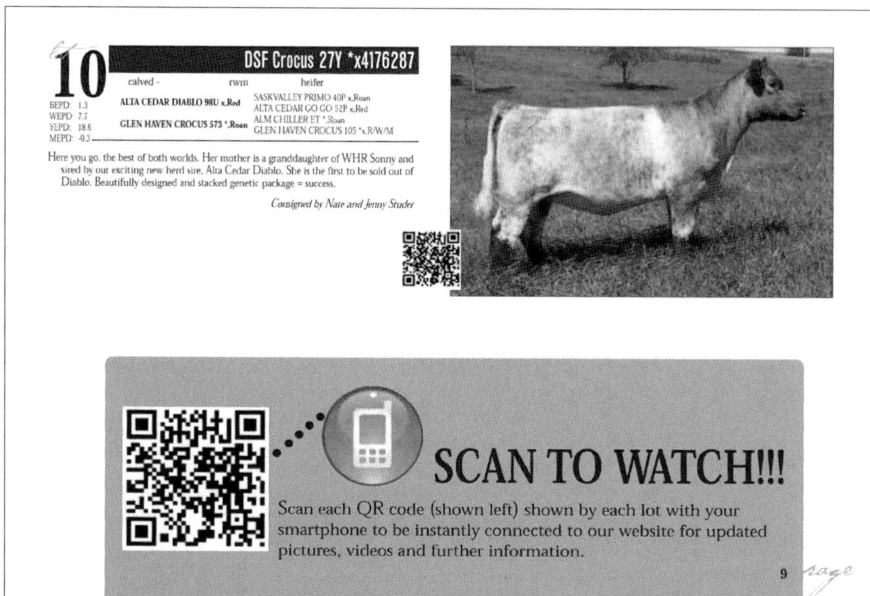

"Family Legacies 2011" by Arin Strasburg for Studer Shorthorns, included QR codes (quick response codes) for each lot offered in the sale. This new feature began appearing in sale catalogs in fall 2011. Used correctly, a reader may scan the QR code with their mobile device and be connected to a mobile-optimized version of the clients website.

Ink preferences are a huge factor in calculating printing costs, but there are several options with varying price ranges. The most affordable option is a catalog printed in black and white. For slightly more money, a marketer can use a black and white catalog with the covers printed in color. Finally, the most expensive option is a full color catalog.

Printers will use terminology such as four-over-four (4/4), four-over-one (4/1), or four-over-zero (4/0) when quoting a catalog. The term 4/4 indicates four-color (full color) printing on both sides of the paper. When four-color is used on the front with black and white on the back it is considered 4/1. Four-color on the front with a blank back is considered 4/0.

The quantity to be printed will also determine the price of a catalog. However, smaller quantities do not always mean a cheaper price. Printers will usually have a minimum threshold quantity that must be met in order to order a print job, usually somewhere around 500 pieces. As quantity increases, the price per unit decreases. There is no magic number to use when determining how many catalogs to order. It is always better to err on the side of caution and have extras available. Having to reprint a small number at a later date will cost more money per catalog. Remember to print enough catalogs to cover a mail out, unexpected requests and to meet event day needs.

Due to increased production and postage costs and the availability of the internet, fewer catalogs are printed than in previous years. This trend is best utilized by printing a small quantity of catalogs for mailing to former and prospective buyers, and enough for those in attendance at the actual sale. The remainder of interested viewers are then directed to the company website, where the catalog can be viewed and printed free of charge to all involved. This strategy often utilizes an email blast once the catalog is available and posted online. The email blast would include an invitation to view the catalog and preferably additional features like sale videos; thus reducing the wait time for printing and shipping.

Once printed, the pages of a catalog must be bound together. Saddle stitching and perfect binding are the most common processes by which this is done. With saddle stitching, the pages are bound by staples on the inside center of the pages, creating a fold in the middle of the document where the pages meet. The perfect binding process involves gluing the pages, creating a spine that is flat and thick. Perfect binding is seen in larger magazines, generally reserved for printed materials over 100 pages where the spine of the publication is greater than .25 inches. Perfect binding certainly increases the quality and appearance of printed material.

Upgrades

Additional premium upgrades can greatly enhance the aesthetic appeal of sale catalogs. These include, but are not limited to, premium printing, premium UV coating, metallic colors, special and translucent papers, textured covers and much more. Check with printers or designers for the pricing of these different options.

Choosing a unique paper size can also help a catalog stand out from the crowd. Standard sizes are normally 8.5x11 inches or 8.5x5.5 inches. Alternative sizes such as 8x8 inch square or landscaped oriented catalogs create a subtle difference from the commonly sized papers.

The effective use of quick response (QR) codes made its debut into sale catalogs in 2011. Each lot in the Family Legends 2011 sale catalog had its own QR code beside it. When the bar code was scanned using customers' equipped cell phones, viewers were directed to a video clip of the corresponding animal. Quick response bar codes can hold various types of information, including video footage.

SOURCES:
The Guerrilla Marketing Handbook
By Jay Levinson & Seth Godin – www.gmarketing.com
ISBN 0-395-70013-2
Published 1992 by Houghton Mifflin Company

Donna Tinsley, Tops Printing, Personal interview
www.topsprinting.com

CHAPTER 5:

Livestock Photography and Video

Just as any professional photographer works hard to create that unique masterpiece, a good farm or ranch photographer will spend the same amount of effort creating the perfect "pasture-piece" image. Today, livestock producers have more options than ever when it comes to hiring professional photographers or handling their own imagery needs through utilizing digital cameras and computer photo editing software. This chapter addresses the basics of good photography, which ultimately helps to increase the value and exposure of a product.

While photographs of animals are imperative for both current and historical reference, many producers are also adding video to their toolbox of promotion. Video gives a much greater visual impact compared to photographs. Additionally, video is perceived as a more realistic image of the animal, since tools like Photoshop® cannot be used to alter an animals conformation in video.

The usage of video in agricultural merchandising is nothing new, however modern technology has made it easier than ever for producers to utilize digital video and thus offer customers a 'real life' visual of their products. In the 1980s, VHS videos were used to promote agricultural products. Many livestock producers hosting annual sales videoed sale offerings, often providing commentary as they filmed, and duplicated VHS copies for potential clients. With the creation of digital video, the entire process has been completely revamped. Although it was once the rare occasion to find a video clip of any type of agricultural product online, today, almost all marketers are utilizing some form of this technique; either through posting video clips online, distributing CDs with digital clips, or both.

The extremely popular use of both photography and video help showcase today's livestock in a very true picture. This allows for viewers and potential buyers to get a visual appraisal of the animal from the convenience of a computer screen.

Film, Digital, and Video Cameras

Though hiring professional photographers has its advantages, many producers also enjoy the opportunity to capture some of their own images personally. There also are times when a photograph is needed at a moments notice, and time may not allow for hiring and scheduling a professional. Many livestock producers, who do hire professionals to shoot key photographs for advertising or sale catalogs, prefer to practice their own skills throughout the remainder of the year by taking all other photos necessary. Either way, the right equipment is needed.

The recipe for success in livestock photography includes a good camera + good natural light + good background + good posing. Fancy equipment is not necessary, but may have some advantages.

Before heading to the store to make a camera purchase, modern merchandisers should consider exactly how the camera will be used. Marketing photos are predominantly used in printed advertising, sale catalogs, on websites, Facebook®, emails, and even as text messages through mobile phone devices. Since each of these mediums requires a different format and resolution, it is best to think about the intended use of the pictures to be taken.

Starting with a good camera is imperative, especially when photos are to be used in printed advertising, as the resolution required is higher than that of online advertising. It is often said, 'the best camera is the one that you have with you,' and that holds true in livestock photography. As today's camera quality continues to increase, sometimes the only camera you have with you is your camera phone, and in that case, that's the best camera you have available.

The disadvantage to using an phone camera is that it produces small photos, poor zoom, and little control over exposure.

If a camera phone is the only camera you have available, shoot with HDR selected. Touch the screen on darkest part of the photo on black animals or lightest part of photo on white animals to help the phone find the right exposure. Apps such as VSCOCam and PicTapGo can aid in photo editing from a camera phone.

Point and shoot cameras are a mid-level camera that will suffice for a lot of producers. These cameras are handy, the mega pixels are very comparable to DSLR, and the attached lenses may have image stabilization technology (IS on Canon, VR on Nikon) to help with camera shake. The disadvantage to using a point and shoot camera is that they offer limited flexibility in focal length due to one fixed lens on camera. A budget of $300 to $500 will yield a good camera for the average farmer or rancher. It is also useful to purchase a memory card with at least 2 GB of storage so you can shoot the largest size images for multiple animals. These cameras are readily available at box retailers.

A DSLR camera is the highest-end camera one can buy for livestock photography.

A DSLR camera is the highest-end camera one can buy for livestock photography. These are substantial cameras with many lens and flash options available. The disadvantage is that these cameras come at a higher price point, and more equipment is needed. Often the "kit" lenses included with purchases are generally lower quality, thus additional lenses are needed.

Before purchasing a DSLR camera, the livestock photographer should choose their preferred brand (Canon or Nikon) and understand that lenses and accessories are not exchangeable between brands. Buy with confidence from camera-specialist stores such as Adorama.com, where one can call and talk to the salesman before purchasing a camera. These sales associates are experts in cameras and will help you determine the best camera for your needs.

Photographers can also consider buying (with confidence) used equipment from reputable camera stores such as Adorama or B&H Photo, where the used cameras are checked and serviced before selling. One might also consider purchasing a camera "body only" and buying lenses separate. "Kit" lenses that come with cameras are okay general-purpose lenses, but are not made for livestock photography.

When purchasing any camera, opt for one with the highest number of mega-pixels within the allowable budget. A camera of at least 6.0 mega pixels will suffice, but the greater the mega pixels, the higher the quality of the photo produced. Most professional photographers chose either Nikon® or Canon® digital cameras with an average of 18.0 mega pixels.

Photographers can conduct pre-purchase comparisons of cameras at www.snapsort.com. Luke and Catherine Neumayr, of Luke and Cat Photography recommended the following cameras by skill/budget:
- Beginners: Canon Digital Rebel T3i (~$600) or Nikon D5100 (~$800)
- Medium: Canon 70D or Nikon D7100 (both ~$1200)
- Advanced Users: Canon 5D Mark III (~$3200) or Nikon D800 (~$3000). Note that these are professional level cameras and not recommended if you are not comfortable shooting in manual or a full-time photographer.

Do you need an expensive camera? As camera price increases, in general, the cameras may have:
- Improved autofocusing systems, which are useful in low light or for tracking moving subjects.
- Better ISO capabilities for low light scenarios such as dark show rings/arenas.
- Higher mega pixels for printing larger images.

In some scenarios these items matter, but in general, livestock photography is done in natural light, so these advantages won't be as noticeable.

Lenses

DLSR's use removable lenses that can be purchased separately. After choosing your camera body, the next step is to choose your lenses. These attachments increase the focal length capability of a camera, allowing it to shoot quality photographs from greater distances. This can be of benefit when picturing animals in large pastures, where the photographer may not be able to get as close to the animals as ideally preferred.

Focal length is the term that refers to how close you are to your subject when you look through the lens. The larger the number for focal length, the more 'zoomed in' your subject will be when you look through the lens. A wide angle lens is 35 mm and below; a standard lens is 35 mm to 70 mm; and a telephoto lens is 70 mm and above.

For livestock photography, the diverse focal length works in a number of different situations. Common and useful zoom lenses for livestock photographers are 24-70 mm f/2.8 (Canon or Nikon), 24-105 mm f/4 (Canon), 70-200 mm f/2.8 (Canon or Nikon).

When budgeting, lens quality is more important than camera quailty. Depending on the lens capability, prices can range from $100 to $2500.

Flash or No Flash?

Shooting in natural light is always a better option than using a flash. However, in dimly lit areas, a good flash is also helpful in creating a more professional photo, and aids in highlighting a subject's natural appearance and details. Most all cameras come equipped with an on-camera pop-up flash, however this is a low quality flash. One alternative to on-camera pop up flash is Speedlight flash that mounts on your camera, available for purchase as an accessory. For most livestock photography scenarios, you do not need a flash. Flashes are most useful for indoor photography needs (show rings), but not needed if well lit.

Resolution

Resolution is the key factor that determines whether or not a photo can be used across different advertising platforms. Print advertising requires photos to be a minimum of 300 dots per inch (dpi), and generally in the format of JPG or TIFF. These photographs are of higher quality due to the increased dpi number, ensuring a clearer, sharper end product. Conversely, a photo taken with a cell phone camera is lower resolution, and could never be used effectively at a large size in a print ad. While you can always downsize an image size and resolution, you can never increase it.

Automatic vs. Manual

Professional cameras offer the choice of shooting in automatic or manual settings. Automatic settings are adequate for users up to a medium skill level. This mode is usually characterized by a green square on your camera dial. In this mode, your camera will automatically set what it believes to be the correct settings for ISO, shutter speed, and aperture. This mode is acceptable for shooting with a good camera in basic lighting situations like natural light outdoors or well-lit show rings.

The disadvantage to shooting in automatic is that the camera relies strictly on what it "thinks" is the scene. This can cause potential problems in outdoor covered show rings where there is a mixture of natural light vs. a shady, darker show ring. In this scenario, the camera may not be able to identify which exposure is your desired one. If shooting in automatic, a basic photo-editing program may be needed to make minor tweaks to exposure (brightness).

When training new photographers, Luke and Catherine Neumayr of Luke and Cat Photography recommend shooting in aperture priority, which is a medium skill level. To use this setting, the user selects the AV (Canon) or A (Nikon) mode. The user sets the aperture and ISO, and the camera automatically sets shutter speed to reach proper exposure. The Neumayrs refer to this as a "semi-auto" mode. In this mode, the user can set the camera to low aperture to help blur background and focus attention on subject animal. This mode offers dramatic improvement compared to auto mode. The Neumayrs state that most competent photographers shoot in Aperture Priority mode every day.

Manual photography is for the total experts, and is where the user sets ISO, shutter speed, and aperture to reach proper exposure. This requires practice and an advanced understanding of what each setting does. Once very comfortable shooting in Aperture Priority mode, users may consider moving to manual if they want more control over images in camera. This is generally used by professional photographers only.

Understanding your camera is important to all photographers. Shutter speed, aperture, ISO work together to control the most basic element of photography: Exposure.

Shutter speed is the click that is heard when taking a picture. This sound represents the shutter opening and closing. The longer the shutter is open, the brighter the image.

Aperture (called f-stop) is the circular hole through which light passes to the camera's sensor. This controls depth of field.

ISO (pronounced as 3 letters I-S-O) refers to the camera sensor's sensitivity to light. Normal livestock shooting is in daylight scenarios (low ISO) or well lit, so this is something most livestock photographers rarely alter. A low ISO would be 100; a high ISO would be 800+. Higher ISO comes at a high price – noise – which is undesirable. Adjusting this setting is usually a photographer's last resort to brighten a photo.

The Neumayr's recommend *Understanding Shutter Speed* by Bryan Peterson and *Understanding Exposure* by Bryan Peterson for further reading.

Additional equipment

Before a photo shoot, livestock photographers should be prepared for various scenarios and make sure they are well equipped with all necessary items. In addition to the camera body and lenses, it is recommended to carry additional charged camera batteries, battery chargers, batteries if needed for flashes, and memory cards. Professionals may choose to carry a back-up camera body, Speedlight, noisemakers or attention getting tools, memory card wallets, and an air blaster for cleaning.

Video cameras

While a film or digital camera is a basic necessity for livestock producers, advanced producers also consider purchasing their own video camera to complete video filming of their animals as well. However, it is often more economical and convenient to hire a professional video service to complete the videoing. Professionals are well equipped with the highest quality equipment, understand editing, and can offer turnkey service that would be very time consuming for producers to attempt on their own. It is our recommendation to clients at Ranch House Designs, Inc. to hire a professional video team.

The process of selecting a video camera can be a challenging decision for amateur videographers (which most livestock producers would fall into this category). Regardless of the intended use of video footage, it is important to consider the zooming capability of a camera. Longtime cattle promoter,

Brad Hook, states that "Zoom is the most critical camera capability where video quality is concerned, especially when filming in a large open area, such as a pen or pasture. However, a high definition (HD) camera may not be required, depending on the intended use of the videos." While HD video supplies a better viewing experience, it also creates a large file size, which many recipients cannot effectively download or receive through email. Straight digital cameras are a safe bet for short clips and personal use. Digital video camera pricing generally falls in the $200 to $2000 range, depending on the quality.

Just as any modern business owner would carefully consider the pros and cons of purchasing a new truck, tractor, herd sire or a new livestock trailer, adding a new camera to the marketing toolbox should be carefully evaluated. Procuring the highest quality camera affordable is recommended; a good camera is often an investment that lasts several years with the proper care and maintenance. In many cases, older cameras, both film and digital, can be easily upgraded by purchasing additional lenses with greater capabilities or specialty features. Determine how images will be used, set a budget, and then consult a retail professional or experienced photographer for assistance in making the best choice.

The Right Set-Up / Picture Pen

Once a producer has their equipment selected, the next step is to create an efficient photography environment. Creating a picture-pen for photography purposes can greatly improve this process, especially when shooting large groups of livestock. For example, a cattle producer may choose to set up a picture-pen directly outside their show barn. This would allow for the breeder to wash, clip and blow out animals immediately prior to turning them out into the picture-pen for that special photo.

The picture pen should be planted with grass, which creates a good foundation for the animal. During the summer months or drought, producers can water the small picture pen area so that a green grassy foundation is always available.

If taking pictures in a pasture setting, allow time for the animals to adjust to your presence. Initially, animals may move away from a photographer. Patience is the key here, as livestock will eventually settle down and allow for the intrusion. Luring animals with feed or hay is not recommended;

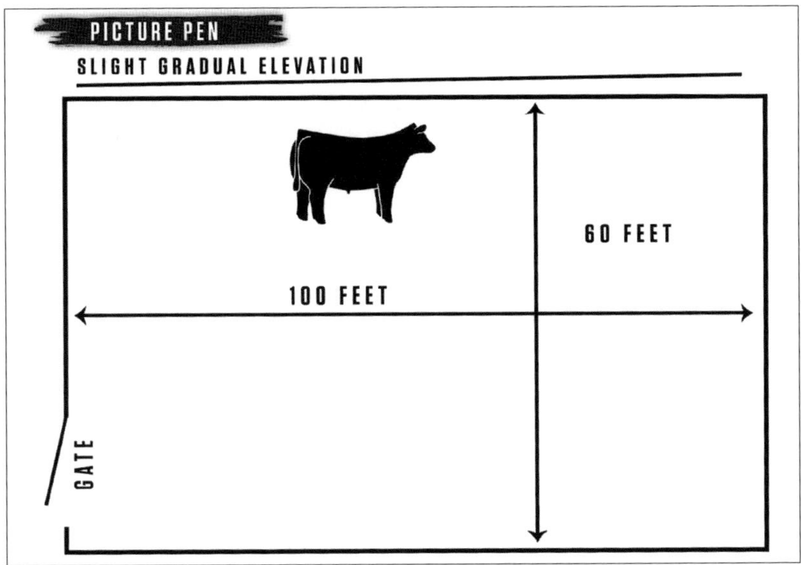

ultimately the result will have the animals focusing on eating, rather than posing. If possible, start visiting the animals in the pasture about a week prior to when you want to take the picture to allow them to get used to your presence on days other than the given photo day.

Prior to taking the photo, take notice of the frame background. While some alterations can be done using photo editing, avoid having distracting elements behind a subject. Ideally the background should be a simple pasture setting. Avoid telephone poles, cluttered fence lines and equipment in backgrounds. Also, take into consideration the color of the subject you are picturing. A contrasting background makes an object stand out. For example, never shoot a Hampshire boar next to a solid black barn, or a Yorkshire gilt in front of a white barn. Some of the most beautiful livestock images capture an animal against a vibrant blue sky while standing on lush green grass, or sometimes clean shavings or straw. Avoid photographing in dirt lots when at all possible.

Zoom and crop are two important elements of photography. While cropping can be adjusted through photo editing software, a good photographer will fill as much of the frame as possible with the subject, yet still allow ample space on all sides for sky and grass. Do not zoom in too close however, as you may inadvertently cut off the subject's hooves or nose.

Timing and sunlight are important considerations before attempting a photo shoot. The best results are usually achieved in early morning or late afternoon, when the sun is not quite as bright and there are not as many shadows. However, depending on the time of year and the location, the hours of 10:00 am and 2:00 pm hold the most fill light. The sun is at its highest during these hours and not near as much consideration has to be given to the subject and photographer's relation to the sun. Keep in mind to always shoot with the sun at your back. Pay attention to where shadows may fall in the photograph; avoid allowing shadows to overcast a subject. It is equally important to take into consideration the time of year. Summer, during mid-day in the South, is extremely hot and may cause heat stress on the photographer and animal. On the other hand winter mornings and evenings in the North are bitter cold, and in extreme cases, may cause equipment freezing.

The Animal's Position

The ideal livestock photo is based on a combination of good lighting, good background, correct body and feet placement for the animal (with all four feet visible), a straight topline, and correct head carriage.

When shooting livestock, regardless of the setting, the ideal position for an animal is to have its front feet positioned uphill from the back feet, with all four feet visible and on a level surface. Picturing animals at halter will allow for a slight mound of dirt to be formed, creating an incline for the placement of front feet. The front feet should be squarely placed in line with the shoulder, with both feet showing if only slightly. The back feet should be scissored, with the photographer's side back foot slightly extended.

When photographing cattle, both ears of the animal should be facing forward, making the animal appear in an alert but docile position. Producers have different preferences for head placement. Some prefer the head to be in a straight-forward position, while others like the head slightly tilted and looking towards the camera. Some breeds, like Brahman influenced cattle who have larger ears, prefer both ears of the animal to be showing.

Typical livestock photos display a side profile view of an animal, however it is becoming increasingly popular to offer a variety of angles, such as three-quarter views or front views as additional images.

Heat Wave, often described as the ideal club calf bull photo, by Christy Collins. All four feet are visible and the animal's head is straight.

Mr. V8 259/7, by Brandon Cutrer. Brahman breeders prefer for both ears of the animal to be visible to show breed character.

When photographing swine, if possible, showmen, judges and friends should kneel behind the animal. This allows for a better zoom and makes the animal seem larger, as shown above. When showmen and friends stand, the photo cannot be cropped as tightly and takes the focus away from the animal, as it seems smaller in the photo below. Images courtesy Duelm's Prevailing Genetics.

Walks Alone, by Chris Rosa. Above, the standard advertising photo. Below, an alternative view. Both images are used in advertising this sire by Rodgers Cattle Company.

The Photography Team and Photographer's Position

An often overlooked, but extremely important factor is to have an effective photography team, which usually includes the attention getter, the pusher, and a handler if necessary.

The attention getter is crucial to ear placement and keeping animals alert during a shoot. There are a variety of tools available to help these assistants do their job, from mirrors and noisemakers to pom-poms and much more. Many of the most experienced in this field simply make animal noises, whistle or move their arms up and down. A good attention-getter attempts to grab the subject's attention before the handler places the feet; it is far easier to set up an animal that is concentrating intently on something else, as opposed to one that meanders into position.

The pusher is the person who moves the animal for the photographer. If your team is limited, the photographer sometimes has to double as the pusher. Sorting sticks or show sticks are helpful in gently pushing animals. The best pusher is one that understands animal behavior and handling. They should first determine the flight zone of each animal, then work to gently move the animal in slow movement to assist the photographer. Horses may also be used on animals used to being worked by horseback rather than on foot.

If photographing animals at halter, a handler is needed. The handler will set up the animal based on the photographer's instruction. As a photographer, remember to instruct the handler to step as far out of the camera frame as possible, as he will typically be edited out of the photo. When photographing sheep, it is acceptable to have the handler remain in the photo. When photographing haltered cattle, another gentle animal can be a very useful tool to serve as a buddy. The companion keeps the subject company and draws their attention as well.

The photographer should be positioned at the center, or slightly off the hip of the animal. Photographers should have their camera at or below the animal's belly. This angle makes the animal appear larger and proportionate; thus the reason most livestock photographers shoot photos while on their knees. Place yourself at the center to the hip of the animal, never towards the front. Back up at least ten feet, allowing the entire subject into the frame and highlighting outstanding qualities more effectively.

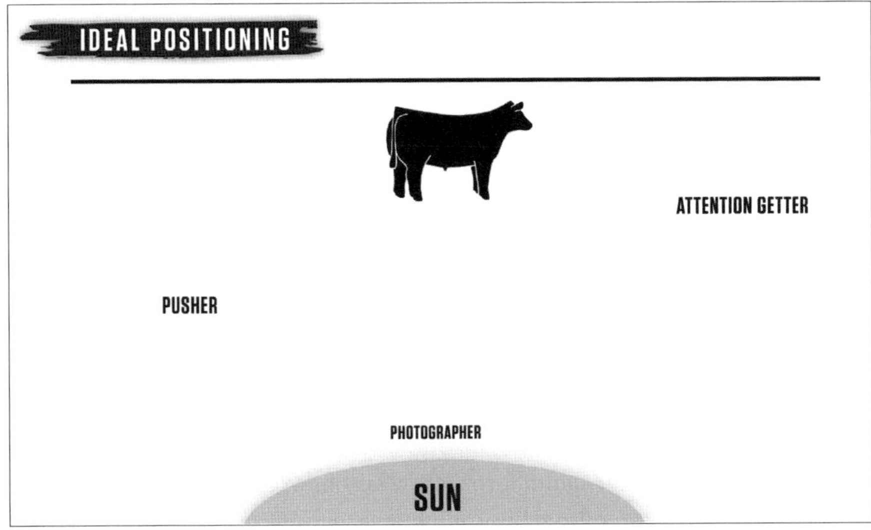

This diagram shows the typical positioning of a photography team.

Photographers bend at the knee or waist to bring their camera to the appropriate shooting level. If necessary, a photographer will do anything to get the shot, including laying on the pasture like veteran photographer Alan Browarny.

Preparation and File Organization

Keep a camera with you at all times, because you never know when you will find that 'perfect pose' for an animal. Producers should constantly be building their image database as the opportunities arise to capture photos of your animals. It is also a good idea to prepare for upcoming sale seasons by scheduling certain 'picture days' in which you try to capture as many photos as possible.

Similarly, once you start capturing your livestock images, it is important to build an organized, consistent photo database on your computer. To accomplish this task, begin as you mean to go. Always put your livestock photos in the same file location on your computer, such as "My Pictures" in your document location. Be as organized as possible. You may choose to create subfolders within your main picture folder such as males, females, show animals or sale animals.

Always rename your image files with the appropriate name of the animal. When naming your files, include as much information as possible, such as tag, sire, dam, and date the photo was taken. This will make your images easily searchable within your computer. It also gives your prospective clients a description of the photo when you email it to them.

An easy to find file name   *A difficult to find file name*
Tag123-Monopoly-Oct2013.jpg   DSC-1234.jpg

Those familiar with Photoshop® may also choose to maintain two separate image files for each photo: a high resolution photo and a lower resolution (smaller file sized) photo that is used for the purpose of emailing to prospective clients.

Building an Image Database

While it is extremely important to maintain an image database of your most prominent animals, sale animals, and show champions, it is also a good idea to build an image database of stock photography that can be used in various promotional pieces. These photos make a great tool when a producer needs a quick photograph for a sale catalog cover, a brochure, or website. Keep your camera with you at all times and be on the lookout for possible photo opportunities such as pasture sunsets, newborn calves, hay meadows, farm signs, personnel photos, show ring shots, and more.

If you capture some of these special moments, consider offering the

It is a good idea to build a image database for your operation that includes scenic photos and special moments like the 'champion slap' pictured here. Photo by Joelynn Rathmann.

photos to the media or your breed associations to use to help in the publicity and advocacy of animal agriculture. The opportunity to have a photograph from your ranch featured on the cover in your breed publication or as an editorial in a trade magazine is worth ten times more than the publicity you gain by purchasing a paid advertisement.

If you are setting a goal of having a magazine cover shot, remember that vertical photos usually work better than horizontal photos. Include lots of sky or grass so that designers can have space for the magazine name and copy.

Special Considerations for Video Editing

Simply purchasing a camera and shooting video footage is merely just the beginning to creating a customer friendly product. Once the film has been captured there are necessary technical considerations and skills required to achieve them.

Good editing software is required to format videos in the proper file type, thus ensuring accessibility across all video player platforms and download speeds. Aside from the software, knowledge of the proper techniques used to compress videos is also important in creating a finished product that achieves the desired results.

File size must be taken into consideration when creating digital videos to be played online. There is a fine line regarding the ideal video size; it must be large enough to maintain high quality, yet small enough to play quickly. Every added amenity placed in a video, such as background music, captions, animations and fancy transitions, increases size. Because large files require more time to download this is a major consideration. As discussed in previous chapters, many agriculturalists are forced to access the internet through slow connection speeds due to their rural locations.

Large video files can also take up huge amounts of space on website file servers. Most

Videos can be hosted on YouTube at no cost. Many professional video firms, like Cattle In Motion (below), offer built-in video players to host client videos.

professional firms will host videos on their servers, providing clients with a link that can be embedded into any website. With this in mind, having the hosting burden removed is an incentive for hiring a professional. Before hiring a videographer, look through other videos online and pay attention to things you like and do not like. Always ask for references and an estimate for the cost of video production and travel expenses. Inform the videographer up-front of any specific deadlines, allowing them to plan far enough in advance to allow for necessary production time. While it is possible for producers to create their own videos, it is often easier and more affordable to hire a professional.

Posting and hosting videos on YouTube© is another option offering multiple incentives. First, YouTube.com hosts videos free of charge and allows them to be embedded into a customer's website. Second, YouTube can provide additional marketing exposure due to the fact that the video cannot only be accessed through the individual's website, but easily searched for via YouTube.com. Advanced users can create YouTube channels for all of their videos, recruit followers and expand their potential client base.

Conclusion

The importance of high quality photography and videography can never be underestimated. Even with the increasing popularity of video, there will always be a need for agricultural photography to be used in printed advertising, sale catalogs, on websites and for historical references. However, video is here to stay, and a tool that will only increase in popularity in the future.

The age old debate continues of whether a producer should take their own photos and video, or hire a professional. Too many times breeders think; "how hard can it be to take a picture or video an animal". These two are best left to professionals who have the equipment and know how to perform these services.

A practical recommendation is to hire a professional for the most important photographs and video such as annual sale photos, herd sire videos, and donor pictures. For the day to day photography needs, a producer should invest in a high quality camera and have the ability to capture these images on their own.

Regardless of the size or goal of your operation, much greater results will be archived by spending the time required to capture high quality photographs of your livestock. Lasting, beautiful images are worth the effort.

SOURCES

Alan Browarny, personal interview, Show Champions - *www.showchampions.com*
The Communicator's Handbook: 4th Edition. Published by Agricultural
    Communicators in Education Organization. Edited by Michael B. Allen & Sarah
    Webb Miller. Published 2000 by Agricultural Communication in Education
Carlos Guerra, Tim Lockhart, Joelynn Rathmann, personal interviews
Catherine & Luke Neumayr, personal interview, Luke and Cat, *www.lukeandcat.com*

CHAPTER 6:

Livestock Web Design

Websites, one of the most affordable advertising options, offer the flexibility of unlimited updates, unlimited photographs and pose as a company's storefront to the rest of the world. While once thought to be a premium advertising option, websites are now a basic necessity. In fact, it is recommended that livestock producers first establish a web presence before participating in other forms of print advertising. The website should serve as a centralized clearinghouse for all information related to your farm or ranch. While a print ad is limited by page size and space, a website has no limitations. Therefore the goal of all printed advertising should be to drive viewers to your website, where they can gain a wealth of information about your business. Implementing a website as part of an advertising campaign is often much easier than expected when following some basic guidelines.

Determine the Type and Kind

Today's agri-businessmen have the option of building their own websites or hiring a professional to create and maintain their sites. The do-it-yourself website is a viable option for those with advanced computer skills. In this scenario purchasing and using the Adobe Creative Suite® software program is recommended. For photo resizing the software package includes Photoshop®; and for html development and content building Dreamweaver® is included. A copy of Acrobat® is convenient when needing to create PDF files. While it is a do-able option, in the long run, most producers choose to hire a professional firm to handle their website due to the affordability and convenience.

Ten years ago, your standard ranch website was done in the html platform, and everyone viewed your website on their desktop computer. Today, potential customers view your website on their smartphones, devices, tablets, laptops, and desktops.

Responsive web design is one approach to website design that allows for the website to automatically respond to the user's device, screen size, and orientation. This approach uses a mixture of flexible design grids and images - often with a more modular or boxy look - to adjust the website to the user. Content management systems like Wordpress® or Joomla offer the technology to create responsive sites, but are more expensive than html. They also offer substantially more advantages including search engine optimization, mobile optimization, thousands of available plug-ins and more.

As of February 2014, Ranch House Designs maintained a total of 571 livestock industry websites. Of these, 95% were built in the html platform. Most mainstream businesses utilize a responsive web design, but this technology has been slow to catch on in the livestock sector.

Blogs are a tool that is very common in the livestock sector. Free programs like Blogger® offer web-based templates that allow users to upload text and photos with different interfaces. The most popular example of such a website is *Matt Lautner's News from the Road* at *www.mattlautnercattle.com*. This blog is updated by the Lautner team anywhere from 10 to 50 times per day. For business owners such as this, the advantage to a blogger-based website is the ability to update easily from cell phone text messages and iPads®.

A common misconception among livestock producers is that a simple blog can act as your sole business website. While this may work for Lautner, it will most likely be insufficient for most producers, who most likely do not have the time to devote to the frequently needed updates. Similar to social media, blog posts are constantly replaced by the newest post, and thus do not provide a good historical online reference. A blog can be used as a nice supplemental tool to your website, but cannot take the place of a website.

Working with a professional firm

For the producer desiring a web presence, yet lacking the time or ability to accomplish it in-house, working with a professional designer is a convenient option. This allows owners and their staff to focus on more important tasks in which they are more qualified. Attempting to learn new software programs and juggle existing responsibilities can often be overwhelming. Professional web design is extremely affordable with many talented design agencies offering a turnkey service of design, hosting and maintenance.

Before beginning a website project, consider the goals the site will be expected to accomplish. It will be important to determine if the site will be updated frequently, or only during key times of the year. The primary products and services promoted online should be taken into consideration. Lastly, decide how visitors of the site will be directed to contact the business. If emails will not be responded to quickly, do not include an email address. Any telephone numbers listed should connect potential clients directly to a knowledgeable person.

Choosing a Domain Name

Choosing a company domain name is one of the early steps of website creation. This name serves as the virtual address of a website with a format of *www.mycompany.com*. The last portion of a domain name identifies the type of website. Commercial sites end with *.com*, educational sites end in *.edu* and the websites of organizations end with *.org*. Domain names should be easy to remember, as specific as possible and closely resemble the business name. The obvious choice is always the exact name of a business. Fortunately for agriculturalists, many domain names are easily available by adding the words cattle, pigs, tractors, hay, etc. after the company name.

Try to keep domain names as short as possible, and avoid using hyphens or other symbols in the name. Businesses with commonly misspelled names should consider purchasing the other spelling variations and having them forward into the original domain address. Professional designers often prefer to register domain names on behalf of their clients, enabling them to handle the technical set up. To check the availability of a domain name visit *www.godaddy.com* or *www.networksolutions.com*.

Web Hosting

Web hosting can be described as leasing virtual space on the World Wide Web. A professional web design agency will handle the technical set up of web hosting in most cases. Under other circumstances, those interested in self registration can check with local internet providers or computer services. An unnumbered amount of web hosting providers can be found on the internet (i.e. Go Daddy®, Network Solutions®).

Developing Content

The next step in creating a website is determining the content to be included. Begin by establishing the specific pages visitors to the site should be able to click on. Jot down ideas of things you would like potential viewers to learn about your business by visiting your website. Most livestock producers choose to include information on both their livestock as well as their business, and even family history. Company history, products for sale, awards and accomplishments, and products of notoriety are a good place to start. Deciding on pages gives navigational structure to a website. Below are common pages among livestock producers:

- Sales
- Winners
- History
- Contact Us

- Photo Gallery
- Sires
- Donors
- News

Once the pages to be included have been determined, begin organizing the content. Work page by page gathering the pictures, captions and any other text. Search engines prefer websites rich with text, consequently placing such sites higher in their ranking.

When developing site navigation and content, strive to convey the intended message in the most convenient way possible. Website viewers tend to become frustrated if more than three clicks are required to find what they are looking for. This rule is especially important for agriculturalists, as rural internet users are often limited by slower web connections. Cut the clutter and make it as easy as possible for people to find key messages. Never use splash pages or introductory pages similar to "Click Here to Enter," which only prolongs the time required for a viewer to reach the desired message. Likewise, certainly do not expose an entirely blank website full of pages that read "Under Construction - Coming Soon." Never design in frames and avoid Java Scripts such as drop down menus that are considered search engine turnoffs.

Content is the key in making a great website, and we recommend these five tips to achieve interesting and relevant content:

1. **Add value and satisfaction to your content.** Have relevant information people want to see. This includes current EPD values, updated show results, news, and other interesting information. Try a video blog. Feature tips, advice and general interest content to entertain your customers. Frequent updates are key to interesting and relevant content.

2. **Always have something listed for sale.** Surfing the web is the most common activity among consumers who use their computers for leisure, according to Advertising Age. When people are browsing online, ultimately they are looking to buy, or conducting pre-purchase research. The livestock producer should always have something listed for sale, even if it is just a general description of private treaty offerings, like *"Pigs are for sale at the farm this April, call us to schedule an appointment."* If you do not have anything for sale, show pictures of past animals you have sold along with a thank you to the

buyers. It is so much easier for someone to look at your cattle for sale online, compared to getting in their car and driving to your ranch. Use your sale page to show prospective clients what you have to offer and build their interest in your program.

3. Tell YOUR story. A great website shows people who you are and what you are about. It helps build a personal connection and shows people that you are worthy of their patronage. Include photos of yourself, your family, and key staff members. To take this up a notch, show a professional video explaining your business history and breeding philosophy. Talk about how you got started in the business, your mission, and what makes you different. Most consumers use the internet for pre-purchase research to determine if a business is worthy of their patronage. Use your website content to reflect your personality and let your potential customers get to know you before they visit your farm or ranch.

4. Get customers to your farm or ranch. The ultimate job of your website is to help you make sales, and in many cases, a face to face connection with your client can help seal the deal. Use numerous calls to action to get people to your ranch...where they are more likely to engage in your operation and make a purchase. On every page, close your content with an invitation like "We welcome visitors to the ranch. Please call us to schedule an appointment." Include your physical address so customers can find you through GPS, or include a map to the farm.

5. Integrate social media. Every business should have a website and a Facebook page. The website is designed for permanent information that can be preserved. The Facebook page is designed for news, humor, ranch life, and continual updates. Make sure that your website and social media are linked so that viewers can easily access both.

Consideration should also be made to determine the mobile accessibility of your website. As more viewers are accessing websites on their phones or tablets, websites should be designed to move fluidly across different mediums. Some tablets, for instance, are not equipped to display Flash® players or certain Java Scripts. While many web users still access websites from a traditional desktop or laptop, increasing number of viewers are using other devices. At this time, recommendations are still to design primarily for desktop and laptop viewing, but offer a mobile optimized version of the site for those viewers browsing on their phones and tablets. All photographs should be optimized for web browsing and ideally all pages should load in under 3 seconds.

Updating

Once the web design is complete, it is imperative to update the site as often as possible. A properly maintained website is updated at least once a month; the best sites are updated weekly. Potential customers are much more likely to frequent a website when they can be assured to find something new every time they visit. Search engines also favor sites that are more frequently updated to those that contain outdated information.

Marketers utilizing print media can benefit greatly from forming the habit of making a website update each time they send a monthly ad to their regular designer or publication. Just as important is regularly reviewing a site to remove any outdated material.

Writing for Websites to Boost Search Engine Optimization

Writing online content for websites is distinctively different than writing for print mediums. The primary difference is that internet users do not actually read entire documents or pages, they scan instead. This scanning is done in just a few minutes, as opposed to the hours readers will dedicate to their favorite periodicals. With this in mind, modern agricultural merchandisers can implement the following strategies to help improve the effectiveness of their websites.

Research shows that website users scan web pages looking for calls to action, specific keywords, photographs or links. Considering the internet is a linking medium, this comes as no surprise. Interestingly, only about 15% of website users actually read web pages word for word; the remaining 85% merely skim. Studies also indicate that emails and email blasts are actually read even faster than web pages.

In order to get the most out of a website, marketers should continually evaluate their web page text. Keep it simple, use key phrases and avoid meaningless copy that only takes up space. However, there is a fine line between having too much text and not enough text, since search engines use the text from websites to determine position in their indexes.

Avoid clutter words similar to "Welcome to our new and improved website, we hope you enjoy visiting," as they do not actually impart pertinent information. Also, never use such empty phrases as "The Best Website on the Internet" or "The Hottest New Website on the Internet."

When composing captions or any other text for an online site, including as many key phrases as possible will encourage visits from prospective customers. For example, an excellent site introduction would read something like this:

When writing for websites, try to include at least 200 words per page for search engine optimization. Visitors tend to skim website text and seek out photographs.

"Welcome to R.A. Brown Ranch, a family-owned Red Angus operation located in Throckmorton, TX. Our annual bull sale is held each October. We also breed and raise quarter horses to be auctioned each fall through our sale."

Notice the sentence includes the business name and location, while giving the reader a clear idea of what the ranch has to offer. It sets the stage, informing users as to what the site is all about. The introduction also includes many key words for search engines to pick up on.

Taking into consideration the scanning process, a good internet copy writer will use simple ways to draw a visitor's eye to the most important phrases. Easy ways to accomplish this are by simply including the most important phrases in bold, or possibly a different color type. Creating links can also serve as a way to highlight important details and engage readers. Subheadings throughout a page are useful as well, giving site visitors a clear idea of how the page is organized. When using large blocks of text seems unavoidable, use bullet or numbered lists which are easier to scan.

Eye tracking research reveals that website readers tend to scan pages in an "F" formation, meaning they start at the top of the page reading the first few lines, then move down the page slightly and read across in a second horizontal movement. Finally, website users scan the remainder of the page following the left margin. This is distinctly different than the Z-format of reading performed in printed advertisements. Successful website writers will place main points at the top of a page (at least in the first two paragraphs) and then be as concise as possible, often using bullets along the left side of the page to illustrate those points.

Experienced news writers use the inverted pyramid style of writing, which essentially starts with the main idea or summary first, then follows with the supporting information in order from most important to least important. This style is contrary to many educational styles of writing where an introduction is used first, followed by three main points, and then a conclusion providing the most important findings. Using the inverted pyramid is more important than ever when writing for website content. It ensures that your readers get the most important information, even if they stop reading midway through the page.

While internet users do not actually read websites, even fewer take the time to scroll down a page. This means that on most occasions a site visitor may only read the initial portion of content visible without having to scroll. Research proves that only those interested in the specific product being promoted will actually take the time to scroll and read through an entire site. Therefore, creating as much interest as possible within the initial viewable section of a website, hopefully enticing viewers to continue reading, is of optimum importance.

Distinguishing it even further from print writing, web text is often more casual. This is especially true for blogs, emails and social media. While the basics of good grammar, spelling and punctuation should always be followed, readers tend to be more lenient on the structure and styles used online.

Accurately citing or referencing sources of information used is just as important to web writing as it is to print writing. When using content from another site, always request permission first, and provide a link directly to that website. Credibility is only enhanced when importance is placed on this matter.

A good rule of thumb is to keep web writing to at least half the word count of that found in print writing, and try to have at least 200 words of text on each page. Remember the basics of the inverted pyramid, using calls to action, and organizing material in such a way as to make it easy for website visitors to learn about the advertised product. Most importantly, remember to include viable contact information.

Conclusion

While once thought as a premium service, an agricultural business website is a modern and affordable marketing tool that should be utilized by all producers. Whether your goal for a ranch website includes something as simple as a static brochure site, or includes a robust, dynamic site where viewers can shop online, a web site is a must. Similar to the debate of whether to hire a professional photographer to take your photos, many wonder if it is possible to create their own website. In both cases, it is our recommendation that ranchers hire a professional for these technical needs, and devote their time instead to the things they understand, like breeding and managing their livestock. The cost of hiring a professional is very minimal compared to the costs associated with purchasing the required software, learning the programs, and troubleshooting when things go wrong. There are numerous professional livestock web design firms that are affordable and trained in creating effective websites. Use them.

SOURCES:
Seth Alling, personal interview
Words that Sell, By Richard Bayan
ISBN 0-8092-4799-2, Published 1984 by Contemporary Books

Eye Tracking Web Usability, By Jakob Nielsen & Kara Pernice
ISBN 978-0321-49836-6, Published 2010 by Nielsen Norman Group

Full Frontal PR
By Richard Laermer with Michael Prichinello
ISBN 1-57660-099-8, Published 2003 by Bloomberg Press

CHAPTER 7:

Signage for Livestock Producers

Signage are necessary components of a promotional plan for livestock producers, ranging from on-premise displays and show animal exhibition, to banners and major trade show displays. Just as retail stores use window displays as their storefront to the world, the agricultural producer's storefront is often a pasture, stall or booth. The importance of professional advertising in these areas cannot be stressed enough, as they are often the image of a business that people remember most.

Signs are a very necessary piece of the advertising puzzle, since a well identified farm sign helps mark your location to clients. Whether your operation is located along a major highway or out in the boondocks, it is definitely to your benefit to display signage. Signage offers many benefits including:

- Providing a billboard to every person passing by a facility
- Assisting visitors in pinpointing a location
- Creating a local market venue, as they inform neighbors of the production of a local company

Farm sign at V8 Ranch entrance.

Barn banners depicting the farm's state fair and major show winners in use at Tusa Show Cattle, Reagan, Texas.

The Farm Sign

The farm or ranch sign should be clearly visible from the road, making it easy for interested parties to see an entrance from several feet away. It should include the business name, logo, contact information and specific products offered. This is especially important in rural areas or desolate areas where it may be more difficult to locate an entrance. There is nothing worse than a potential buyer arriving at your place of business upset and frustrated because he spent the last 30 minutes trying to find you.

Typical business signs include something similar to: *"V8 Ranch – Registered Brahman Cattle"* and a phone number or website address. An arrow pointing in the direction visitors should go can also be very helpful. Even better, include an eye catching photo or drawing of one of your best animals to showcase the type and quality of the livestock you have on your operation.

Local sign makers are not as common as they once were, but some do still remain. It is worth checking with available local companies for pricing because they can often design, construct and install custom signs.

Purebred livestock breeders can check with breed associations or state associations to see if they offer a standard member sign. Many times these organizations offer template designs, with minor personalization to allow for a company name, at very affordable prices. National Cattlemen's Beef Association, Farm Bureau, Texas and Southwestern Cattle Raisers Association and other organizations offer smaller signage for gate attachment to help identify ownership of property.

Sign vendors are prominent at the trade show portions of major stock shows. Besides distributing their contact information for future use, often these vendors provide customized and affordable signs on-site, in a short turn-a-round. At the very least, they will take orders and ship at a later date.

Vinyl Banners

When company signage is needed, but time does not allow for completion of a permanent design, vinyl banners can be utilized temporarily for special occasions. A few large vinyl banners, with general wording such as "Welcome to Topline Farms," hung from entrances or gates can help direct visitors to a location for field days or sales. After the event, the banners can be easily removed and stored for future use.

Barn banners are also a great use of signage as a promotional tool. Tusa

Thomas Charolais billboard, located along one of the major highways in south Texas. Left: Vehicle lettering and trailer lettering serve as a moving billboard for your ranch.

Show Cattle, for example, purchases a custom barn banner for every steer they sell that wins a championship at a state fair. These banners are hung in the barn in clear eyesight of all visitors (and prospective buyers) who visit their farm. This is a great way to let buyers see your record of success.

Billboards

Though not as common as farm signs, billboards offer a promotional opportunity to livestock producers owning road frontage property. Billboards fall under different jurisdiction than typical company signs, and may require special permits or approval, depending on location. If your operation resides on a major highway, take advantage of this prime real estate. A billboard along a major highway reaches hundreds of thousands of viewers.

Decals and Lettering

Do not overlook the importance of using your farm and ranch vehicles for promotion and advertising. Every piece of equipment should be customized with the business name, and logo if possible. As livestock producers, when we are travelling and pass a livestock trailer on the highway, how many of us wonder who that fellow producer is and if we know them? Likewise, others on the road are drawn in to look at livestock trailers. Invest in the time and expense to have your trailer identified with your name and/or brand on key areas of the trailer such as the nose, back gate, and side of the trailer. Make sure your tack and equipment also displays your name. Showboxes, hog panels, blowers, and more can be easily identified with a quick decal. This not only is a great form of advertising, but also a theft deterrent as well.

Conclusion

Farm signs and other methods of identification are often overlooked, but necessary. Initially, it may seem like a large amount of time, effort, and expense is needed to complete these tasks. However, the effort is worthwhile, and provide long lasting results in your promotional campaign.

CHAPTER 8:

The Role of Livestock Shows as Advertising

The best way to advertise your product is to actually show others examples of your livestock. By participating in competitive livestock shows or trade shows, a livestock producer can showcase their highest quality animals to huge audiences of fellow producers. Despite the expense and time required to exhibit at major livestock shows, many producers feel there is no better way to get your animals in front of a multitude of breeders. Remember, anytime your livestock are on display there is the opportunity for promotion to prospective clients.

A show string or display animal helps promote not only your ranch, but your breed. Shows are the more glamorous part of the purebred livestock industry as they serve as a showplace for livestock types and trends, and often a social gathering of the leading producers in the area. Many times, prospective buyers go to major livestock shows to look for a new herd sire or breeding female. Participating in shows also helps the smaller breeder get his name out to large groups of people. Exhibition at major livestock shows provides the breeder an opportunity to compare their livestock with others to assess their breeding program. It also provides an opportunity for all producers – big or small – to keep their name in the public eye.

There is no doubt that exhibiting at livestock shows requires intense effort and increased expense. One can easily spend thousands of dollars in any given year exhibiting livestock when factors like feed costs, entry fees, equipment, gasoline, hotel expenses, labor, and more are considered. Producers can expect show costs to be anywhere from $10 to $20 per head per day to keep an animal in a show barn, and up to $100-$200 per head per day for taking animals to shows.

Showing is expensive, but what many consider a very necessary part of advertising. To get the most exposure from your showstring, consider the following:

- *Booth, Stall and Pen Signage* – Every operation should have clear

"Ahh, the Show ring...where I can lose old friends and make new enemies." -Dr. Miles McKee. Whether a fan or opponent of showing, the show ring provides an excellent forum for promoting your livestock.

signage identifying the business name. This may include a large vinyl banner, end panel signs, silk banners, company logo decals on equipment, etc. This signage should be large enough for spectators to recognize while walking through aisles or pens.

- *Hospitality or Booth Area* – Depending on the space available, producers may wish to set-up a small booth area. This arrangement can be as simple as a few chairs and a display table, or as elaborate as the mind can imagine. Such an area provides a nice spot for clients to have a seat, relax and hopefully negotiate deals. More times than not this extra space must be purchased from the livestock shows (entering extra animals to plan for space) and planned for in advance, but a justifiable investment considering it provides comfortable area for potential customers. If facility regulations allow, drinks or snacks offer a nice touch as well.

- *Individual Animal Signage* – Every animal in your stalls or pens should be clearly identified with a smaller signs stating the animal name, date of birth, pedigree, name of owner, sale lot number when applicable and any other pertinent information. This helps let the animals "sell themselves" by providing useful information to spectators.

- *Personalized Ranch Apparel* – All associated staff and employees should be dressed professionally and in company attire. Caps, jackets and shirts, labeled with the company logo, give an added level of professionalism. Remind anyone wearing your ranch apparel of that while wearing your logo, they are a representative of your business. Any expectations such as no tobacco or alcoholic beverages allowed while in company attire should be clearly communicated. Also, let your team members know if certain clients or individuals are expected to visit the booth area so they may be greeted warmly and made to feel welcome. Brief employees on key points of certain animals or products prospective customers may be interested in.

In general, all staff of your show string should be instructed to greet every visitor with respect and hospitality. Provide as much information as possible when questions are asked, and make personal contact with those who take time to stop and visit your booth.

Never let an interested spectator leave your booth empty handed. Offer

End panel signs, hung at the edges of your stall area, help spectators identify your location.

Booth spaces can range from a simple arrangement of tables and chairs to exquisite artistic displays depending on the level of the show.

a business card, a flyer, or a brochure to any visitors as a reference to your operation.

Special Promotional Displays

There are certain occasions that call for special individual promotional displays, such as Denver bull promotions "in the yards," boar displays at the World Pork Expo or special displays of prominent animals at a sale. These arrangements require a bit more planning and creativity to help draw attention to the special product.

Generally, such exhibits include a large single area, dedicated to one individual product. Pen sizes and shapes can vary, depending on the animals being exhibited. The allotted space should allow enough room to showcase the animal in clear sight and allow the animal enough room to move freely.

Anytime an animal is showcased in this manner, prepare far enough in advance to have the animal presented at its best. Display animals should be in good condition, professional clipped or groomed, and clean at all times. You never get a second chance to make a first impression.

Visual aspects of displays such as this usually include one large banner featuring the animal's name, photo, pedigree and owners. Pen banners for cattle are usually 4x4 feet if two animals share the same pen and 4x8 or 5x10 feet for single animal displays. Banners for featured show pigs may be smaller, due to smaller pen size, but also range from 2x2 to 4x4 feet.

Owners get elaborate on these displays, vying for the crowd's attention. Some additional finishing touches that may entice viewers to an area include:
- Specialty colored shavings
- Theme displays utilizing a décor that complements the name of the animal or product
- Showing video footage of the animal on a large screen television
- Background music, or a live band in the display area
- Providing food and drink hospitality
- Offering free promotional give-a-ways (i.e. koozies, caps, etc.)

While bells and whistles may aid in creating crowd hype, ultimately the quality of the livestock is what spectators want to see, so make sure the animal remains the focus of the display and is clearly visible at all times. Remember that phenotypic quality is the first thing that draws buyers to an animal. Never let anything detract from showcasing the animal.

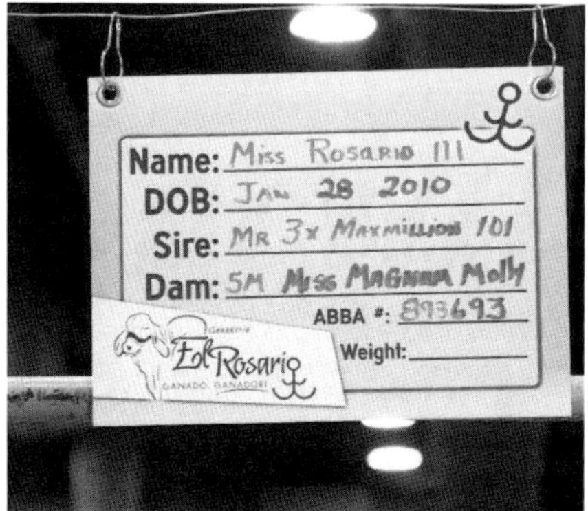

Every animal exhibited in a show string should be identified with an individual sign stating their name and pertinent information. Permanent signs for each animal range from $20 to $50 per head. Dry erase signs are also an option for those ranches who would like to re-use the signs through the years.

Special pens can be set up at shows and sales to draw attention to important reference animals. or sale features.

Are Shows and Outdoor Displays Worth It?

Participating in shows and events is a much debated role in the purebred livestock business. Some can't live without going to shows, others wouldn't enter a show ring if their life depended on it. However, the show ring is not for everyone, especially those operations whose advertising or expense budget simply cannot handle this expense.

Whether you're a proponent or opponent of shows, one has to agree that showcasing animals in outdoor displays like major livestock shows bring much needed attention to the livestock business, breeds, and individual breeders.

The thrill of competition is a huge factor in drawing crowds to major livestock shows. However, ultimately, showing is a form of advertising. The reason for going to the time and expense of preparing, feeding, fitting, and exhibiting livestock is to put your product in front of prospective buyers. To do this, you must be present at shows where there are large numbers of people. Your placing (along with your competitions) are publicized in your breed publications, local newspapers, television, and on the Internet. This, along with keeping your name and livestock in front of the public eye, are possibly the most expensive - but most worthwhile means of advertising a producer can use.

SOURCES
Guerrilla Marketing Attack
By Jay Levinson – www.gmarketing.com
ISBN 0-395-47693-3
Published 1989 by Houghton Mifflin Company

Merchandising Bulls at Private Treaty, by Martha A. Hollida, published in The Brangus Journal, September 1982

The Communicator's Handbook: 4th Edition
Published by Agricultural Communicators in Education Organization
Edited by Michael B. Allen & Sarah Webb Miller
Published 2000 by Agricultural Communication in Education

The Show Ring In Today's Cattle Industry. An Interview with Knic Overpeck. Simbrah World, September 1984.

CHAPTER 9:

Social Media for Livestock Producers

Ten years ago, most progressive livestock producers could simply put up a website, and that would be enough to satisfy their online marketing needs. However, today, modern livestock marketers need much more than just a website; they need a web presence.

This is where social media comes into the picture. Social media is one of the newer trends being utilized by modern livestock producers. However, if you spend more time working on the ranch than you do in front of a computer screen, the phrase "social media" may seem like a foreign language.

Wikipedia® defines social media as "media used for social interaction, using web-based technologies to turn communication into interactive dialogues." The online encyclopedia continues by stating that "current marketing trends show the use of social media to be a driving force in blending technology with social interaction to create value for a product."

There are many social media outlets; however the most common ones among the agriculture industry are Facebook®, Twitter®, YouTube®, Pinterest®, Instagram®, Vine®, Snapchat®, and many more. Message boards, forums, and podcasts are also examples of social media. These sources are very different from traditional outlets of public information such as magazines, newspapers, or television, and new social media outlets are continually added or gain in popularity to the point that they become relevant to the livestock producer.

As we continually recommend, every successful modern livestock marketer needs a web site to begin their online presence. The website serves as your permanent online hub for your business, with ultimately all social media efforts being directed back to the web site for complete information.

However, the days of just putting up a web site and then forgetting about it are over. With the trend towards more viral, constantly updated online material, today's producers should also be utilizing some form of social media to take their marketing to another level. Once the foundation of a ranch web site is established, the top marketers will use several other tools to use for your online marketing efforts.

Benefits of Social Media

The two primary advantages to using social media are cost (they are usually free) and user-friendliness. Social media outlets are generally very inexpensive, if not completely free. Users often find they are able to connect with fellow agriculturalists from around the world, with just a few clicks of the mouse, creating a virtual network of friends and colleagues. This helps livestock producers to promote their products on a very large scale, reaching both local and worldwide markets. Users may update their Facebook or Twitter pages in a matter of seconds, using a computer or cell phone.

Social media websites like Facebook® and Twitter® are extremely easy to use and navigate. Only basic computer skills and internet knowledge are required to join conversations and activities, update pictures or share news. Users can update pages in a matter of seconds using a computer, or even a cell phone, through these straight-forward interfaces. For example, a merchandiser wishing to post new pictures on Facebook® simply has to click on the desired photos. The website then automatically uploads, resizes and publishes the photographs.

An added benefit to using social media sites is anything can be edited, changed or even completely removed in a convenient manner. A typographical error found after a sale catalog is printed cannot be changed, whereas typos on the internet can be permanently corrected quickly and easily.

The instantaneous time factor cannot be discounted as a huge benefit to social media marketing. If a producer wishes to get the word out about an upcoming sale, something similar to "New bull sale catalog just posted at www.myranch.com," can be posted as an update to the company media page. This update is then broadcast to all members or friends of the site. No time-consuming phone calls or mail-outs necessary, just a few simple keystrokes will accomplish the goal.

Concerns for Using Social Media

Once you join communities like Facebook or Twitter, chances are you'll be hooked. In 2013, Advertising Age reported that the average consumer spends 40 minutes a day using social media, and 316 minutes a day in total online activities. It's very easy to get caught up on chatting, checking your friends news, and more. Set boundaries and limitations for the time spent on social media, especially for children.

By adjusting privacy settings, a user can control how much information is visible to other site users. For example, a company or personal profile can be made visible to friends only, or made completely visible to anyone in the entire world.

Choose words carefully when interacting with others through social media. Avoid posting controversial or offensive topics that might alienate potential clients. This practice is one that has become an increasing problem in the livestock business, where we frequently see social media users bashing judges, bashing fellow exhibitors, or even bashing fellow breeders. Smart livestock marketers should abstain from being a cyber bully or assuming that you are an authority on a given topic. Instead, remember that you are one of millions of participants in the social media conversation. Make an effort to offer relevant and timely contributions, always being courteous to others.

Along the same lines, make it a goal to post things that are interesting, newsworthy, funny or unique. Don't clutter the online community with irrelevant status updates like "I just ate lunch" or "I'm bored." Even worse, don't overuse social media to try to shove your marketing message down users throats. Just as people hate spam emails or telemarketers, no one likes a person that continually is posting updates that try to sell or over-market their product to their friends. Instead, try to build a reputation as someone who posts useful information that others enjoy – and even look forward to reading.

A Look at Useful Social Media Tools for Ranchers

Current trends in marketing show that many companies are boosting their efforts in social media. In February 2014, more than 400 social media outlets existed worldwide. Alexa's Global Traffic Rank and U.S. Traffic Rank from Compete and Quantcast data show the top ten most popular social networking sites to be Facebook, Twitter, LinkedIn, Pinterest, GooglePlus, Tumblr, Instagram, Flickr, VK, MySpace, and Tagged. How does one even begin to decide which outlets are necessary in their livestock marketing toolbox? It is our recommendation that all livestock producers utilize Facebook, then select possible additional outlets based on their time availability and experience. For most livestock producers, at this time, Facebook is sufficient.

Facebook®

The most used social media outlet worldwide, as of 2013, Facebook maintains more than one billion active users worldwide. Facebook cites that the average user spends approximately 16 hours per month on the platform. Facebook is available on desktops, laptops, smartphones and tablets.

Facebook is a low-cost marketing strategy, free to join, that is ideal for small to medium sized businesses with a smaller marketing budget. Larger businesses can use Facebook to trial marketing concepts before committing to big campaigns. Facebook is great for creating your brand and instilling it into the consumers ideas about you, what your represent, and what you have to offer.

So how does a livestock marketer use Facebook? The main advantage is that this platform allows users to reach a huge potential audience of interested clients. Status updates allow you to provide instant updates on new sale pictures, sale updates, ranch news, and other events. It helps you connect with your audience on a personal level and is easy to use and manage on your own without the need for professional help.

When getting started on Facebook, it is very important to understand the difference between personal and business pages. The personal aspect of Facebook is your Facebook timeline. This is a personal account for individuals, and are intended for non-commercial use. A common problem occurs when businesses attempt to create a personal page for their business, i.e. First name: Ranch House, Last name: Designs. This is a violation of Facebook's terms of use, limits you to 5,000 friends, and offers no analytics or page management functions. On the bright side, if you have made this mistake, perhaps setting up a Facebook account with the first name Smith and last name Club Calves; Facebook can help you convert this personal account to a business page through the "help" tab.

Facebook Pages look similar to personal timelines, but they offer unique tools for connecting people to a topic you care about, like a business, brand, organization or celebrity. They are managed by individuals with personal timeline accounts who have been designated as administrators of the page. A page may have multiple administrators, each with different levels of access for posting or viewing of the page tools.

As a Facebook page administrator, users are provided with a wealth of data insights to help you manage and track your page engagement. There is no limit to the number of users who can like a business page, and users can install an array of apps to enhance your Facebook page.

Prior to using Facebook, it is a good idea for all parties involved in the business to have an open discussion about the responsibilities of Facebook and discuss what type of posts the business will be comfortable posting. Discuss which members of the team will have access to post on Facebook on behalf of the business, and which won't. Discuss the type and kind of posts that the business likes to post, and any topics that they wish to avoid. It is very important to remember that while individuals will be posting on behalf of the business, ultimately all posts made from the business page are a reflection of the business image.

To get started with a Facebook page for business, visit www.facebook.com/business, where you will be walked through a step-by-step process of creating your page. The following checklist is helpful in creating a Facebook page:

1. Identify yourself. Choose your page name, which will be the way that your business will be identified on Facebook, i.e. Ranch House Designs, Inc. is our page name. If your name does not help identify the product you offer, consider including additional words to help tell users about your business. For example, V8 Ranch's Facebook page name is V8 Ranch – Brahman and Shorthorn Cattle, so that our viewers also understand what breeds of cattle we raise. These additional words also help your Facebook page perform better in search engines.

Secondly, choose your custom page address in your settings. When you first join Facebook, your page is assigned a random address with numbers and letters, which is how your page is identified until you set a custom URL. To use a custom page address, a page must have 25 or more followers. Custom URLs make it easier for people to find your page and to list your Facebook address in printed material. Example: www.Facebook.com/v8ranch.

Also, be sure to complete information about your business under your page information. Include your address, phone numbers, website, date founded, mission, history, products, awards, and as much information that you have available on your business.

2. Look professional. When setting up your Facebook page, you have the option of uploading two standard images that serve as your thumbnail and your cover photo. Typically, the business logo can serve as your thumbnail image. The cover photo can be an eye appealing photo or a professionally designed graphic. While the thumbnail photo will remain constant usually, the cover photo can be changed out frequently to showcase events, feature animals, or provide updated material for your fans.

3. Engage and recruit fans. Once your page is established, and graphics are in place, it's time to start connecting with fans. Conversely to personal accounts where users must confirm their friends, any user on Facebook can like your business page without approval or confirmation. The goal is to build a steady Facebook fan following of other Facebook users who are interested in your product.

Make sure that your website has a link to your Facebook page. Invite your friends to like your page. Send an email blast to your contacts to promote your page and get likes. Like other relevant pages or run a contest to help build your following without investing in paid advertising.

Aside from the free methods of building a fan base, investing in paid Facebook advertising is the #1 best practice for growing your following. The Facebook advertising platform offers immense targeting options, flexible budgets, result tracking, and analytics to measure the return on investment of your ad.

4. Plan and manage your content. When people like your page, they are saying that they care about your business and want to know more. Posting relevant content is the most important thing you can do to keep them interested.

The recipe for quality Facebook posts include short, visual posts optimized for key time periods. Posts between 100 and 250 characters receive approximately 60 percent more likes than those posts who's text overflows into the "continue reading" option. Photo albums, pictures and videos get 180 percent, 120 percent, and 100 percent more engagement respectively compared to text only posts.

Use a variation of content strategies to keep you content fresh and interesting. Carefully plan content that includes a mixture of promotion, sales, and non-sales related material. Here are a list of good content generating ideas:

- Post copies of your print ads on Facebook when they are approved to give viewers a sneak peak of what will be published in the coming months.
- Post an "animal a day" from your sale lots
- Choose a themed day of the week, i.e. Motivational Monday, Ag Fact Friday, Throwback Thursday
- Give an inside look to your business through casual pictures taken on your camera phone of daily activities like newborn animals, weaning & weighing, your display at stock shows, etc.
- Don't be afraid to show your own picture! People love seeing their friends, or their friends children. Ranch kids are always a hit.

A good Facebook marketer will post status updates a least 1 to 2 times a week, often more. This help your information stay top-of-mind to other users and provide relevant information. Facebook pages offers a very powerful scheduling tool which allows you to preparing posts in advance and scheduling when they'll show on your page.

5. Use Insights. Page Insights is a free analytics tools for your Page. Insights are used to help you understand your audience and the content to which they respond. It provides very important demographic information including the age, sex, and location of your fans. By studying insights, users can identity the type of posts that are most shared and talked about, and create more posts of this type more frequently. These insights help you better understand your customer base.

Perhaps the most compelling reason for the success of livestock advertising on Facebook is the simple fact that it is fun! When managing a Facebook page, don't stress out or be overwhelmed. Much of successful Facebook marketing comes from trial and error, so have fun, experiment, and find what works best for your business.

Will Facebook help you sell your livestock? Yes.

Twitter®

Different from Facebook, Twitter engages users through short, snappy status updates of 140 characters or less. These status updates are called tweets. This is the second most popular social media outlet, and one that livestock producers should consider. Launched in 2006, Twitter has more than 500 million active users who post an average of 340 million tweets a day. This platform is available on desktops, laptops, smartphones and tablets.

This platform allows users to either follow, or be followed, by just about anyone in the world. It helps you engage with users you most likely do not know, and may never know personally. By interacting with other users, you can re-tweet their posts, reply to their posts, or favorite their posts to show engagement. Twitter is a convenient marketing tool that is simple to use, mobile friendly, and easy to update on-the-go.

When using Twitter, it is a best practice to realize that Twitter is not Facebook. Posts should be short, newsworthy, and more frequent. The best Twitter users post at least once a day, often multiple times daily. Space our your tweets and post during the key times of 8 to 9 a.m.; 2 to 5 p.m.. Research shows the best retweet rates between noon and 4 p.m.

Twitter users have a unique vocabulary of terms that may sound like gibberish to a novice social media user. Here is a breakdown of popular Twitter terminology:

Tweet - a post on Twitter, 140 characters or less.

Twitter-handle: the name you choose to represent you or your business on Twitter. Once you pick it, it will be shown with the "@" symbol in front. Example: @ranchhouseinc

Follower: someone who has chosen to receive your Tweets

ReTweet (RT): reposting what another user has posted to Twitter. Usually denoted with "RT @username".

Hashtag: A way to tag a keyword or topic within a post to give it emphasis and make it more easily searchable by other users. A hashtag is notated with the "#" symbol.

Hash tags are used to search through other users' posts to follow popular topics. For example, a search using the "#cattle" hash tag would return the most recent Twitter posts containing the term "#cattle." When posting, include appropriate hashtags to help others find your content, such as your breed of cattle (#brahman, #angus), #stockshowlife for junior shows, #ranchlife for ranches.

Hashtags are also very common at major national events to help others stay connected with the events activities. For example, the National Western uses the hashtag #nwss. During the champion drive at the junior market steer show, those attending the show will likely be posting tweets about what is happening. A Twitter user who is at home could easily search #nwss and view a live feed of all the tweets relating to the Denver stock show at that time.

So how can Twitter help a livestock producer? It allows you to connect with a broad audience through hashtags and groups, broadcast information and updates from events, and stay informed about events.

Will Twitter help you sell your livestock? Maybe.

LinkedIn®

While Facebook and Twitter are social networks designed for casual interaction and personal connection, LinkedIn is the world's largest professional network with millions of members, according to the LinkedIn corporate profile. LinkedIn is a social tool that allows you to post a professional profile, similar to a very advanced, detailed online resume. On LinkedIn, you connect with other professionals and colleagues in the business. LinkedIn offers an outstanding job posting tool that allows companies to search, target and recruit qualified applicants.

How can LinkedIn help livestock producers? While it is an extremely relevant and useful network for professionals, there is minimal benefit for a farm or ranch to being on LinkedIn other than to create a company profile and recruit employees.

Will LinkedIn help you sell your livestock? No, but it will help you find qualified job applicants and allow you to connect with other colleagues in the business.

Pinterest®

Founded in 2010, Pinterest is a social tool for collecting and organizing the things that inspire you, according to www.pinterest.com. Designed as a virtual bulletin board, Pinterest users *Pin*, or bookmark, items that they like to virtual bulletin boards. Pinterest users can pin items from websites, blogs and other Pinterest users boards. The Pin is archived on the users boards and will always link back to the site it was originally pinned from.

Mashable reports that nearly 70 percent of Pinterest users are female, with 28 percent having an annual household income of more than $100,000. Nordstrom, Whole Foods, and West Elm are the brands with the largest Pinterest followings. Most pins on Pinterest are related to fashion, beauty, home decorating, or food.

Will Pinterest help you sell your livestock? No, but it can help build agricultural awareness and share recipes. It is also extremely useful for agri-businesses that offer consumer goods such as clothing, food, or other retail products.

Instagram®

Launched in 2010, Instagram is a social network that is designed exclusively for photos. As of 2013, it maintains more than 100 million active users and is only available on mobile devices: iPhone, iPad, iPod Touch, and Android. It is not available on desktops or laptops.

Instagram is based on the basic premise that pictures speak louder than words. Instagram users, including businesses, use their camera phone to create a window to their world.

How does a livestock marketer use Instagram? First, create and consistently use a company hashtag, like #breedlautner or #v8ranch. Include this hashtag on relevant images, and encourage others to do the same. This will make it easier for people to find your brand, whether they use Instagram or not.

Second, Instagram is a much more personal network compared to others. It gives your audience a behind the scenes, personal glimpse into your business. Images posted on Instagram can be very casual, very down to earth, and this makes people feel that your business is "real"and approachable.

Instagram offers fun photo filter effects that turn your images into artistic works. The filter gallery can help you create black and white images, sepia images, vintage photos and more.

By using Instagram, users can build brand awareness, building relationships with a teenage and college age crowd, and showcase the agricultural lifestyle. Will Instagram help you sell your livestock? Probably not.

Participation in the Online Community

Finally, participation in the online community is imperative to building an online presence. Today's market is all about social interaction. By hitting the "like" button in Facebook®, or re-tweeting a pertinent statement on Twitter®, a business can vastly increase its exposure. Posting comments on favorite blogs and joining message board discussions are great ways to interact with business associates, potential customers and other industry professionals. When posting or commenting, provide feedback that is relevant, polite (even if you disagree), and not self-promotion.

Blogging, Facebook®, and Twitter, are just a few of the thousands of ways you can build your online presence. We recommend livestock producers set a goal of participating in one or two of these activities, as time allows. A good starting point is to set aside 15 minutes each day to participate in social media efforts. These modern livestock marketing tools will help boost your ranch's exposure and name recognition.

SOURCES:
Ashley Culpepper, personal interview
www.Facebook.com
www.Twitter.com
www.wikipedia.org

CHAPTER 10:

Email Marketing for Livestock Producers

Not so many years ago, producers hosting annual sales were forced to complete their catalogs nearly a month in advance of the sale date, wait a week for the printing process to be completed, and yet another week for the catalogs to be stamped, shipped and delivered. Today however, livestock producers can approve the final copy of a sale catalog at noon, and have it email blasted to thousands of potential buyers in 15 minutes or less through the powerful tool of email marketing.

As one of the newer technological advancements available to modern merchandisers, email blasts are continuing to increase in popularity. Usually containing the most updated news stories or advertisements, these email messages are delivered to large groups of subscribers. Many livestock producers utilize email blasts over printed or mailed newsletters because of the convenience, cost-savings compared to printing and postage, and instantaneous delivery.

Building a List

The first step in beginning any email marketing campaign is to build an email address list. While it may be tempting to skim through publications and type in any and every email address available, this is actually a huge faux pas for an email marketer. Your mailing list should be permission-based, meaning that subscribers have voluntarily joined by providing their email addresses. It is also important that anyone on an email list be able to voluntarily unsubscribe at any time.

How do you begin to assemble an email list? To initiate the process, modern merchandisers should have a place on their website and Facebook® page where users can subscribe to the mailing list. Customers and registered sale buyers can be asked to provide email addresses, for the specific purpose of being added to a mailing list, at the time of doing business or registering for an auction number. Trade shows and stock shows are also opportune places to collect email addresses by requesting that booth visitors leave their business cards to join lists.

If you wish to create your own email blasts, it's a great idea to sign up for a web-based email marketing provider such as Constant Contact. This is a fee-based subscription service that allows you to create email blasts using their templates, and manage your mailing list online. Constant Contact helps you make sure your emails meet anti-spam requirements, permission-based requirements, and provides real-time tracking and reports to help you know how many people read your emails and clicked on your web site. Their process also makes it quick and easy for users to subscribe or unsubscribe from your emailing list, which is very important.

Using a Professional Service

For those without an existing address collection, there are many third-party lists for sale through breed associations, sale managers and advertising agencies. These industry professionals will often assist in creating and designing email blasts as well.

The price of email blast services, just like printed media, is largely based on the lists circulation and demographics. Email blast service providers should be able to tell prospective clients the number of subscribers on the list as well as the target market of the lists. Before using a third-party firm to send an email blast, ask the firm the following questions:

- How many good, valid addresses are on their list?
- What are the demographics of the list?
- How current is the list, and how often is the list updated?
- What options are available for email blast reporting, such as click through rates?

After asking these questions, producers should choose one or two email blast service provides that best fit their needs. For example, a purebred Angus breeder wishing to promote his Angus female sale would be best served by purchasing an email blast from the Angus Journal. A cattleman wishing to promote a bull sale to a large, diverse cattle audience would purchase a list from CattleMailUSA. Just as with any service, expect to pay for quality. The cost of purchasing an email blast service may range from a few hundred to a few thousand dollars depending on the quality and circulation of the list.

Designing an Email Blast

The goal of designing email marketing is three-fold: get the viewer to open the blast, click on the blast, and ultimately reach the business website where more information is available. To accomplish this, three tactics are used:

- Write an effective email subject line
- Design an attractive and user-friendly blast
- Include a call to action link to the website

Today's culture is fast paced, and connecting with potential customers through their mobile phones is very important. Blasts should be designed to be mobile friendly, and offer an alternative link for those who may have trouble accessing the blast on their phone.

Dan Leo, CattleMailUSA, was a pioneer in email marketing to the cattle industry. He feels that interactive content is the key to a successful email blast. With the growing number of videos being used to represent sale offerings, an email blast serves as a way to offer 'one-click' access to the clients website, which offers a portal to the photos, videos, and catalog.

As email marketing continues to grow, more research findings are available to indicate the elements that can help attract readers and what elements turn readers off. Most notable among these findings is the annoyance felt from a spammed email. Avoid sending repeated annoying emails to a subscriber list; jokes, forwards or other distractions only aggravate recipients.

Email blasts should include meaningful, worthwhile information only.

Subject lines

The moment of truth for most email blasts is the split second it takes a reader to decide if they will actually open the email or delete it in efforts to keep their inbox free of clutter. This is where the subject line is more important than ever. Use a concise subject line that tells the receiver exactly what the email is concerning, such as "100 Prospects Selling This Thursday." Avoid meaningless subject lines similar to "Don't Miss Out On This," or even worse, the dreaded "Urgent! Reply Immediately." The aforementioned subject lines are often mistaken for spam. The "from" field in the email should contain a name that is recognized by the receiver, such as the ranch name, or the business name.

This email blast consists of a large graphic area to attract readers and is accompanied by plain text which is accessible throughout various mediums. It also includes a large call to action to drive visitors to the website.

Blast content

Research has also proven that users spend less than one minute reading the average email newsletter. Just as readers scan website content, they scan email blasts even faster. With this in mind, it is important to design an email blast that is visually appealing, easy to read and arranged in such a way as to

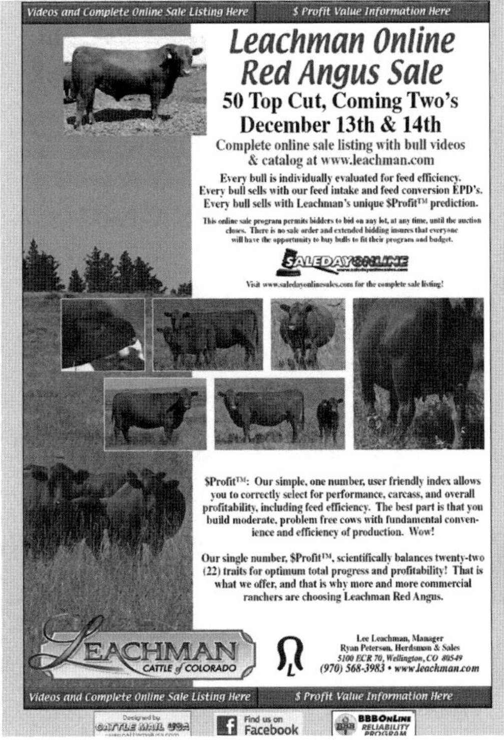

Leachman Online Sale, by CattleMailUSA. This email blast consists outstanding use of design and calls to action to attract the viewer to visit the website for sale videos and more information. It also encourages social media participation through Facebook.

encourage readers to take time and read the message, rather than quickly hitting the delete button. Take into consideration that many subscribers are accessing their email from mobile devices, so make it as convenient as possible for these users by including a mobile link in any blasted email.

Good email newsletters or advertisements will be brief in content, use legible fonts, and provide worthwhile information. Remember that time is valuable for readers, making the effort to provide a user-friendly, fast loading email blast that is relevant to target readers of utmost importance.

Additionally, email blasts provide a way to support and update a customers previous print advertising campaign. According to Leo, about 75 percent of his customers use updated photos in their email blasts compared to photos that were previously featured in sale advertising, which was due several months earlier to printers. This allows ranchers to showcase a photo of his sale lots in their most current shape. Or, ranches may blast updated breeding information or calving information. These updates and photos help maintain clients interest in the product by offering them something new.

If using multiple blasts to advertise the same product, change the blasts slightly for each delivery. Research shows that viewers are more likely to click on something new, opposed to the same exact blast they saw a week ago.

The Call To Action

Every blast should use the call to action strategy to invite viewers to click on the email blasts desired link, where viewers are able to obtain in depth information. For example, if using an email blast to promote a sale, the blast may contain photos of a few key sale features and then a call to action of "Click to view videos of the sale bulls." Upon clicking, the viewer is immediately re-directed to the specific page on the website where the videos are posted.

Evaluating Email Marketing

The success of an email marketing campaign can be easily evaluated based on several criteria. After any blast is sent, merchandisers should measure the following results:

- Number of emails sent, number of emails bounced (invalid addresses)
- Number of subscribers who reported the email as spam or opted-out
- Number of opens
- Number of click-throughs (those who followed the hyper-link)
- Number of forwards
- Number of social media shares

Any invalid addresses should be removed from future mailing lists if unable to verify a correction. Careful notice should be taken on spam reporting or opt-outs, as these are indications that emails are being sent too frequently, or the emails are viewed as boring or distracting to subscribers.

Higher percentage values are desirable in the number of emails opened, number of click-throughs, number of forwards and number of social media shares. The goal of any email blast should be to direct people to a website, via hyper-link, and to create a viral message that can be easily shared on the internet.

Conclusion

The need for printed materials will never lose it's value, however, modern livestock merchandisers are now trending more towards electronic means of delivery compared to printed. For example, a producer who may have printed and mailed 10,000 sale catalogs in 2000 may only print and mail 1,000 catalogs today, specifically targeting his established customer list, having enough catalogs on hand for sale day, and printing a few extra for special phoned in catalog requests. Then, he will purchase an email blast service to broadcast his catalog to the mass public. Currently, the best advertising campaigns involve a strategic mix of printed media, email blasts, and social media that ultimately drives viewers to the business website.

SOURCES:
www.constantcontact.com
Dan Leo, personal interview, www.cattlemailusa.com

Guerrilla Marketing for Free, by Jay Levinson
ISBN 0-618-27679-3
Published 2003 by Jay Conrad Levinson

CHAPTER 11:

Establishing A Budget

When taking a business program to the next level with advertising, regardless of the mediums, there are certain steps one should take to ensure the optimum amount of exposure is obtained for the money spent. The ideal advertising plan includes a calculated mix of purchased, earned and social media. Purchased media is that which is bought and paid for by the advertiser. Paid media include magazine ads, charitable sponsorships, email blasts, and Facebook advertising. Owned media is the media that you control, for example, your website, your sale catalogs, your social networking statuses. Earned media is free publicity obtained through news releases, editorial content, viral social network shares, word of mouth, your show ring accomplishments, and

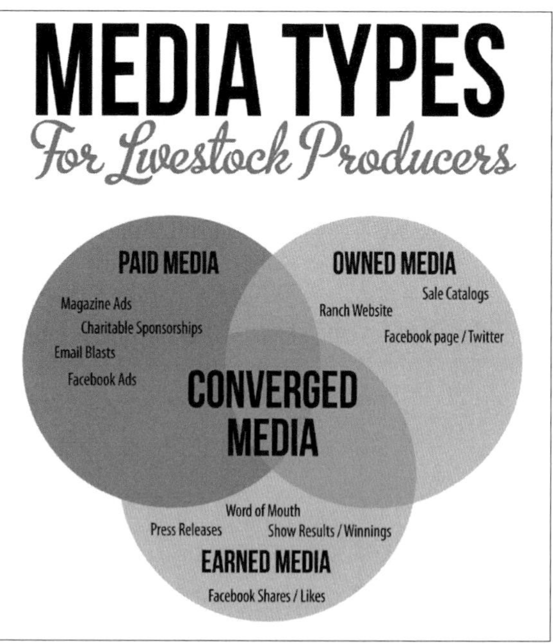

public relations. All of these types of media work together to create the total media presence for your business.

By carefully choosing the desirable publications with which to advertise, using specific strategies to get higher reader response and supplementing with free publicity and exposure, producers can maximize exposure and ultimately income.

Budgeting On A Per Head Basis

One method of calculating an advertisement budget is to budget a certain amount per head that you wish to spend on advertising. Jim Dismukes, Dismukes Ranch, markets the majority of his bulls private treaty and likes to plan for a cost of $150 per head for advertising. To calculate this figure, divide the total amount spent on advertising for the year by the number of cattle sold each year.

Budgeting Based Off Gross Sales

Another method of calculating an advertisement budget is basing it on a percentage of the total sales of a business. While there is no magic number, many agricultural producers set budgets from 1% to 5% of annual gross sales.

Julie French, owner of McMahon-French Promotional Strategy in Beaverton, Michigan, recommends evaluating the goals and budget of each account to maximize the exposure and reach for each client. She conducted the following case studies:

CASE STUDY #1:   East Texas Angus Association Sale
Held annually, first weekend in December
East Texas Angus Association is a regional beef cattle organization that hosts a cooperative bull sale for its members. The target market for this event includes regional purebred and commercial beef producers in the areas of Texas, Louisiana, Oklahoma and Arkansas, with a budget of $1500 to $3000 for the purchase of a bull. French worked extensively with sale managers to create an advertising budget that would effectively promote the sale without racking up a huge expense for participating members. This budget included a mix of smaller sized (1/2 page to 1/4 page ads), black and white print advertisements in magazines and newspapers, and local radio advertising.

Industry Newspaper Advertising

Angus Beef Bulletin	53,000 national circulation (8048 in target market)	1/2 page ad October	$670	Monthly Publication

All Breed Publication Advertising

The Cattleman Magazine	17,000 national circulation	1/4 page ad - November	$240	Monthly Publication
The Louisiana Cattleman	5,300 state circulation	1/2 page ad - November	$295	Monthly Publication
Gulf Coast Cattleman	16,000 Southeastern circulation	1/3 page ad - November	$285	Monthly Publication

Regional Newspaper Advertising

The Weekly Livestock Reporter	10,300 national circulation (4,365 in target market)	Calendar ads - 5 issues prior to sale	$1124	Weekly Publication
		1/2 page ads - 3 issues prior to sale		
Country World	20,000 East Texas circulation	1/4 page ads - 3 issues prior to sale	$506	Weekly Publication

Radio Advertising

KIVY Radio		1 ad - during the ag news report	$50	Week Run

Total Advertising Expense: $3170 $46.61 per lot

Sale Gross: $131,325
Advertising Costs: 2% of sale gross

The advertising definitely performed. The sale sold 68 lots for a gross of $131,325 and average of $1,824. Buyer analysis showed that 53% of the buyers were on the association's catalog list and were mailed a catalog from the association. That leaves 47% of the buyers without a previously mailed catalog; indicating they learned about the sale through the other advertising.

CASE STUDY #2:   **Great Shorthorn Revival Purebred Cattle Sale**
Held annually, third weekend in September
The Great Shorthorn Revival is an annual production sale hosted by Little Cedar Cattle Company in Michigan. The sale allows for consignments from fellow Shorthorn breeders who have purchased genetics from Little Cedar or the Great Shorthorn Revival in the past. Compared to the East Texas Bull Sale, this event targets a nation-wide market of purebred Shorthorn breeders and junior livestock showmen. The sale includes high-end purebred genetics including show prospects, donor females, elite herd sire prospects, semen, embryos and pregnancies. Advertising for this event focused heavily on the Shorthorn and junior livestock market. Minor emphasis was directed toward the regional Midwest market in hopes of attracting local buyers.

The Great Shorthorn Revival advertising included the following publications, newspapers and internet marketing:

Shorthorn Breed Publication Advertising

Shorthorn Country	3,500 national circulation	1 page ad, 4 color - May 1 page ad, 4 color - June 2 page ad, 4 color - July 2 page ad, 4 color - August 1 page ad, 4 color - September 1 page ad, 4 color - November (thank you ad)	$5,800	Monthly Publication

All Breed & Junior Publication Advertising

The Show Circuit	4,300 national circulation	1 page ad, 4 color - August (State Fair Issue)	$500	Monthly Publication
Prime TIME Showman's Handbook	6,700 national circulation	1 page ad, 4 color	$300	Fall Handbook

Regional Newspaper Advertising

The Live Stock Exchange	10,000 national circulation	1/2 page ad, black and white	$100	Monthly Publication

State Publication Advertising

Michigan Cattleman	3,000 Michigan circulation	1/2 page ad, black and white	$180	Monthly Publication

Internet Advertising

Ranch Website		www.redwhiteandroan.com	$450	Full time

Total Advertising Expense:	**$7330**	$103.23 per lot

Sale Gross: $312,325
Advertising Costs: 2% of sale gross

Consignor evaluations and sale results found the advertising to be very effective. The sale sold 71 lots for a gross of $312,325 and average of $4,399. Each consignor paid slightly over $100 for their part of the advertising campaign which reached both a national audience of Shorthorn breeders with buyers in 10 different states as well as local buyers from Michigan. The sale also attracted new buyers, who were not on the catalog mailing list, indicating they learned about the sale through Internet or printed advertising.

CASE STUDY #3:   V8 Ranch Annual Advertising Budget
Year round private treaty sales marketed at the ranch

V8 Ranch is a purebred Brahman and Shorthorn operation in Texas. The Brahman operation sells cattle primarily on a private treaty basis to purebred seedstock producers (domestic and international), junior show cattle exhibitors across the southern United States and commercial bull buyers for crossbreeding programs. The ranch consigns one elite animal per year to the International Brahman Sale and hosts an annual internet sale in September. The Shorthorn segment of the operation markets cattle through the WHR Lone Star Edition Sale every April, as well as private treaty sales throughout the year. Advertising efforts are directed to purebred Shorthorn breeders and junior livestock showmen.

The ranch advertising campaign includes international and national media, promotional give-a-ways, and a heavy mix of internet marketing.

Originally, it relied heavily on print advertising, including at one time holding the back cover of The Brahman Journal for nearly 10 years. However, the ranch as recently reduced their advertising budget by $20,000 per year by eliminating some print advertising during slower months and increasing their social media presence (which had no costs).

Print Publication Advertising   **$21,500**

Australian Brahman News	Australian circulation	Business card ad, black and white	Quarterly Publication
Desorrollo	Central American circulation	1 page ad, 4 color - 4 times per year	Quarterly Publication
Shorthorn Country	3,500 national circulation	1 page ad, 4 color - July (herd reference issue)	Monthly Publication
The Showbox	6000 national circulation	1 page ad, 4 color - February 1 page ad, 4 color - April 1 page ad, 4 color - October	Monthly Publication
Gulf Coast Cattleman	10,000 regional circulation	1/3 page black and white ad in all issues	Monthly Publication
The Cattleman Magazine	17,000 national circulation	1 page ad, 4 color - October (bull buyers guide) Business card ad - 12 times per year	Monthly publication

Digital Advertising   **$3000**

Ranch Website, Email Blasts, Facebook, Twitter, YouTube

Promotional Apparel   **$5000**

Each year the ranch distributes approximately 1000 hats and T-shirts to ranch visitors. These are especially popular with international visitors who treasure mementos from their visits to American ranches.

Total Advertising Expense: $29,500

Gross Annual Sales: $1.2 million

Advertising Costs: 2.0% of sale gross

The ranch has found they could cut some of their printed advertising without sacrificing sales. They use seasonal printed advertising to reach the more established, older clientele who are members of breed associations and currently in the cattle business. They place a high priority on a search engine optimized website for targeted keywords to reach those interested in the cattle business. In fact, when new clients come to the ranch, often they report they found V8 Ranch on Google. While they ranch continues to use print advertising for printed reference and within-breed advertising, they have found that the majority of their new clients are made through contact on the internet.

CASE STUDY #4:   Circle A Angus Ranch Spring Bull Sale
200 head Angus bull sale held in March

Circle A Angus is one of the nations top seedstock and cow/calf operations, located in Missouri. The ranch hosts two sales each year: spring and fall. The spring bull sale features 200 head of Angus bulls and Certified Circle A Premium Bulls along with 200 head of Angus females. The ranch uses a mixture of print publications, email blasts, social networks, radio and web advertising to promote the sale. Of their budget, 88% of their advertising budget is spent on print advertising. They develop 3-4 general sale advertisements that can be placed with different publications throughout the central and southern United States. Print ads are placed in publications in January, February and March (three months preceding sale) in both national beef publications and regional publications in Missouri and surrounding areas. Many of the publications utilize a smaller size ad (half pages and 1/3 pages) in black and white format, so the importance of an outstanding design with a clear message are crucial to the sale's success.

Print Publication Advertising $38,000

National publications: Angus Beef Bulletin, Angus Journal, Progressive Cattleman

Regional Publications: Alabama Cattlemen Magazine, Arkansas Cattle Business, Cattle Business in Mississippi, Cattle Today, Cattlemen's News, Cow Country News, Farm Talk, Gulf Coast Cattleman, High Plains Journal, Midwest Cattleman, Midwest Marketer, Missouri Angus Trails, Missouri Beef Cattleman, Ozark Farm and Neighbor, Tennessee Cattle Business, The

Cattlemen's Advocate, The Stock Exchange, Weekly Livestock Reporter

Digital Advertising: $3500

    Ranch website

    Web banner ad on Beef Magazine Cow/Calf Weekly

    Email blasts through CattleMailUSA, American Angus Association, and Ranch House Designs

    Facebook page with daily updates 1 month prior to sale

Radio Advertising: $900

    Radio spots on KRES, KRMO and KKOW (Missouri and Kansas stations) for two weeks prior to sale.

      Gross Annual Sales: $1.8 million
      Advertising Costs: 2.5% of sale gross

Summary

Every component of an advertising campaign should be evaluated to determine its rate of success. This is not always an easy thing to accomplish, but often the best method for finding out how customers are learning about a business is to simply ask them; a mere "How'd you here about us?" will suffice. If mail-outs are included in an advertising program, compare the annual purchaser list to the mailing list to determine if clients are repeat buyers or new clients. To measure the effectiveness of a sale advertising campaign, analyze the buyer list following the sale, comparing it to the company mailing list and determine if the buyers were current or new customers.

If you are purchasing advertising, and still getting no inquiries, calls, or emails, then this would be an indication that the producer should take a good, hard look at their program. Jim Leachman, of Billings, Montana quoted at one time, "If you don't get inquiries, it is most likely the fault of the program, or the result of unrealistic expectations, or the fact that the ad did not ask for a response. Prospective buyers need something to respond to. Give it to them."

Careful, precise planning and consideration should go into creating an advertising budget. Always include an allowance for those unexpected publication deals that arise throughout the year. As with any budget, once a number has been determined, try to stay within the financial parameters that have been set. For agricultural businesses that do not host production sales, the advertising budget should still be calculated using total gross sales. However, there may be a focus on seasonal or annual advertising when private treaty sales occur, or centered around major events in which the business participates.

Most producers are always looking for ways to reduce expenses on their livestock operations. However, difficult times are not the time to let up on advertising. Good advertising is not an expense, but an investment. Remember, results are not always immediate. The longer a product or company is advertised, the better. A terrible thing happens when you don't advertise: nothing. Ultimately, success in advertising and marketing should be measured by one thing: profitability.

Sources:
Guerrilla Marketing Attack
By Jay Levinson – www.gmarketing.com
ISBN 0-395-47693-3
Published 1989 by Houghton Mifflin Company

Julie McMahon French, McMahon – French Promotional Strategy
Nick Hammett, Circle A Ranch, personal interview

CHAPTER 12:

Sire Promotion Basics

Each year, literally hundreds of new sire prospects emerge as possible candidates for use in artificial insemination programs throughout the livestock industry. Over the past ten years, the sire promotion business has experienced a tremendous growth in popularity, especially among beef cattle and show pig producers. While the idea of owning and promoting that "hot new sire" is appealing to many, the process involves extreme attention to the merchandising process. A successful sire promotion campaign includes broad exposure in print and internet media, as well as tremendous word-of-mouth promotion. In many cases, sire promotion also includes a visual display component, such as publicly displaying the animal at livestock shows or expos to allow the mass public to view the animal.

Start with Quality

Individual quality of the animal is the number one factor that attracts potential buyers to your product. Before making the commitment to begin the intense promotion process of an individual animal, breeders should ask themselves the following questions:

1. Is the animal high enough quality to offer improvement to the industry?
2. What traits does this animal have to appeal to fellow producers? (I.e. performance, outstanding pedigree, calving ease, carcass merit, show ring accomplishments)
3. Are you willing to invest the necessary time and money into an intense sire promotional effort?

While there are hundreds of new sire prospects that emerge each year, the number of sires that become widely used throughout the industry for years at a time is very minimal. Producers should carefully consider the potential for their animal to make a meaningful impact on the industry before committing to the large expense and effort required to enter into a sire promotion.

Choosing a Name

As the first step in this promotion process, owners of promotional sire prospects often deliberate for days, weeks or even months before choosing the official name for their animal. Inspiration is generally drawn from a variety of sources. Trendy names can come from popular movies, musical artists,

The animals name will highly influence the style and design of the promotional advertising for the sire. For example, Wave Goodbye was named in honor and memory of Kent Habeger.

athletes, politicians or songs. Traditional names often include a moderate variation of a name found in the sire's pedigree as a way to draw attention to the bloodline. For example, the sire Heat Seeker inspired the name of his son, Heat Wave, who inspired numerous progeny names based on the use of the word "heat" or other weather-related names.

Purebred cattle breeders, on the other hand, use combination of cow family names, sire bloodlines, and identification numbers to create names, such as SAV 8180 Traveler 004. While the number "6807" may seem irrelevant to most, any Angus breeder in the world could quickly identify this bull as DHD Traveler 6807.

Once a name has been chosen, announcing the new bull via website or Facebook® page will decrease the chances of another animal emerging with the same name. Copyright and trademark issues should be addressed prior to choosing an official name. If there is any question, consult a lawyer regarding the rights to use of the name in a commercial promotion.

Official Photograph and Video

Promotional sires are most often identified by one signature photograph, used as the primary visual representation of the animal throughout its life. In the club calf business, such photos are taken when the animal is a yearling; purebred cattle breeders prefer photos taken at a later stage in an animal's life; and the show pig industry has established six to eight months as the preferred age of promotional sire photos. Regardless of the species or the age, obtaining an outstanding picture of a promotional sire is of the utmost priority. Hiring a professional photographer should definitely be considered, thus ensuring a high quality photo capable of use in both print and web materials.

Many sire promotion specialists also recommend hiring a professional to capture a video of the sire in addition to the photograph. The video should be posted on the business website, used in social media, and available on YouTube. The video is a very useful tool allowing the public to get a visual identity of the animal in motion.

Printed Marketing Materials

Once a name is selected and the official photograph of a sire is secured, it is time to create printed promotional material. The basic format of utilizing a large photo allows potential buyers to gain a clear and precise depiction of the animal. Printed advertising should specifically tell consumers why they should use the sire presented, as opposed to the hundreds available choices. Pertinent information regarding sire, dam, date of birth, genetic testing, weight, awards and successful siblings accomplishes this task nicely. Semen price and ordering information, along with any necessary contact information must be included as well. When selecting graphics or slogans copyright issues should be considered, especially when the animal's name involves common terms or brands in popular culture. It is never acceptable to use a trademarked logo or name without obtaining proper licensing from the trademark owner.

Evaluate potential magazines to determine which publication(s) will offer the best advantages in terms of target audience, advertising costs and circulation. Once a print ad is developed, purchase additional copies of the ad to be posted as flyers at any industry related events.

Displays of Sire Prospects

Many producers choose to display promotional sire prospects at major industry events, allowing the public to evaluate the animal in person. Sire displays are especially common at major livestock shows and related conferences, industry conventions and beef expositions. Each year, the cattle industry's new sires are usually displayed at the National Western in Denver as well as Midwestern beef expos. Boar prospects are displayed at events like World Pork Expo, and National Swine Registry Regional Type Conferences, and Midwestern state fairs.

If participating in a visual sire display, bull owners should present the animal at the highest level of professionalism. The animal should be professionally groomed, clean, and presented on fresh bedding. Signage should prominently display the sires pertinent information. A knowledgeable staff member should be available at all times to answer questions from prospective clients and assist in presenting the animal to its fullest.

For the beef industry, many times a bull may be displayed prior to being collected. Some producers choose to offer a pre-order option or 'waiting list' for those wishing to purchase semen on the bull when it is available.

The typical promotional sire display requires some form of oversized signage to attract clients to the animal. Sire display banners range in size from 4x4 feet to 5x10 feet and are printed on all-weather vinyl. Depending on the nature of the display, promotional give-a-ways, such as caps featuring the

A knowledgeable staff member should be available at the booth or display pen at all times to answer questions from the public and make sure the animal is presented at his best.

sire's name, or other items help build brand recognition for the promotional animal and its owners.

Internet Promotion

All modern sires should have a strong online presence. Begin by featuring sires on a company website or Facebook® page to introduce the animal to the industry. The official photograph of the sire should be published online as soon as possible. Publishing video clips on the internet can also be very advantageous. Consider email marketing as part of the internet promotion, as such marketing is a proven and effective tool for spreading the word about available sires. Email marketing is also useful in announcing to the public when semen is available on sires or to invite the public to view the animal at upcoming displays.

Typical sire banners are 4 x 8 feet, printed on weather resistant vinyl, and mounted on metal or plastic frames. Banners are most effective when they display the sire name, photo, and information.

Teaming Up With Semen Distributors

Collaborating with a professional semen distributor can provide great incentives for sire owners. These distributors can assist in marketing and sales, while providing convenient ordering options for potential clients.

The nature of working with semen distributors may vary between different companies. Some of the larger distributors may choose to purchase exclusive distribution rights to the sire, or purchase the sire in his entirety. Others may elect to purchase volume groups of semen on the sire to re-sell to the public, or accept semen on consignment from the breeder or owners. There are numerous reputable semen distribution companies and individuals available. Inquire with several distributors and develop a professional relationship to aid in the marketing of the sire.

Conclusion

It does not take long to incur a large expense when beginning a sire promotion program, especially if the sire is displayed at an event. Typically, producers do a very heavy push for promotional sires during the first year genetics are on the market. After that, promoters should evaluate the amount of income generated by the sire, and make a decision regarding an increase or decrease in future promotional efforts involving the animal. The modern livestock marketer will carefully plan and budget, evaluating each segment of the promotional campaign to determine its effectiveness.

Sources:
Jon Fisher, Prairie State Semen Service, Personal Interview
Lance Ellsworth, CattleVisions, Personal Interview
Matt Lautner, Personal Interview

CHAPTER 13:

Ethics in Livestock Advertising

Ethics, like integrity, is often described as 'how someone acts when no one is watching.' In advertising, agricultural producers must adhere to certain standards, regulating themselves to tell the truth. While most businesses strive for honesty in advertising, there are some gray areas that are cause for ethical concern.

Advertising Themes for Youth Audiences

In the agricultural industry, the prospective clientele ranges from young children all the way through older adults. Livestock advertisements in particular are often directed toward youth livestock exhibitors. These 4-H and FFA members range in age from 8 to 18 years old. However, children of all ages have access to the livestock magazines in their homes and barns; therefore, all agricultural advertising should be appropriate for audiences of all ages, or in other words, "G-rated."

Unfortunately, in certain factions of our industry there are producers who feel giving animals names with sexual or alcoholic implications will create mass appeal or aid in creating a cool factor to prospective clients. Good marketers will keep in mind that while an inappropriate name may create hype among the 18 to 21 year old crowd, it is more likely to alienate the older, more established breeders who are making more of the purchasing decisions.

Similarly, agriculturalists should exhibit caution with the use of alcohol or tobacco products in advertising, especially if the animals name may be similar to the names of alcoholic beverages . If a questionable name is already in use, take caution when selecting graphics for promotional purposes. If the use of alcohol or tobacco product graphics is unavoidable, at the very least do not use a picture showing people actually partaking in the products.

Product Placement & Endorsements

Product placement is a part of everyday modern advertising. Absolutely no one could miss the glasses of Coca-Cola® placed strategically on the American Idol judges' table during the filming of the show. In the agricultural industry, the use of endorsements is similar to product placement. Commercial company logos are becoming commonplace on the printed advertisements and websites of various producers, insinuating the producers' use and satisfaction with the corresponding products.

At the time of publishing, the Tusa family from Texas, owners of Tusa Show Cattle, secured product endorsements from Showmaster® and Sure Champ™. Of course these two companies were interested in endorsing with Tusa Show Cattle due to the integrity they are known for exhibiting. The logos of Showmaster® and Sure Champ™ have appeared in Tusa magazine ads and on their website. The Tusa's must adhere to certain standards of advertising set by Showmaster® and Sure Champ™, and in return, these companies will pay for portions of the Tusa advertising budget.

When a business accepts endorsements, both parties should take into consideration that their actions are reflective of each other. For this reason, always use caution when accepting or soliciting endorsements. While any way to reduce advertising costs is tempting, affiliations should be chosen wisely.

Ethical Photo Retouching

With the increasing availability of photo editing software, which often comes standard on many new computers today, modern merchandisers face an ethical dilemma of just how far is too far when it comes to photo retouching. To maintain a level of professionalism and respect, every producer should adopt a set of ethical retouching standards.

Ethical photo editing, such as cropping or color correcting, can greatly enhance the aesthetic appeal of photographs. Many times, it is easier to correct an image, than it is to reshoot a product. Image alterations are perfectly acceptable, and quite honestly necessary to achieve the highest quality photograph. However, altering the physical qualities of a product is considered unethical, and should not be practiced.

Ethically cropping an image eliminates empty space or unwanted items from a picture, such as extra sky and grass, unnecessary equipment or structures, telephone poles and the like. Ideally an agricultural product will fill as much of photo frame as possible, leaving an equal amount of background above and below. Cropping also allows the centering of an image within the frame to allow for equal horizontal spacing of the subject. Accomplishing these tasks through cropping helps direct readers' attention to the most important thing in a photo, the animal. It is NEVER acceptable to alter an animal's physical conformation or the physical characteristics of any animal in any way, shape or form.

Color correction can drastically improve images intended for advertising. Very basic, and somewhat limited, color correction can be accomplished by adjusting automatic color levels and contrasting through photo editing software. In Adobe Photoshop® "Auto Levels" or "Auto Contrast" options can be found under the Image Menu of the program. While these are not perfect, they do offer a quick fix for many images. There are also many free color adjustments and Photoshop® actions available on the internet for adjusting image lighting and color. While this may work in some cases, the user is allowed very little control over specific color alterations. A more advanced editor will use manual adjustments of color levels, brightness and contrast to improve the image. Professional or experienced photographers will also find that shooting and working in a camera's raw settings will allow the most flexibility in color and lighting correction.

Removing halters from animals or erasing handler body parts is considered ethical photo retouching. This can be done in a matter of seconds through using the Clone Tool in Photoshop®. Again, the use of this simple technique merely places the primary focus on the subject, and does not alter the physical characteristics.

Ethical retouching can also be used to gently clean up animals or backgrounds. Livestock producers are forever fighting the weather and many times this fight takes place on the day of scheduled photo shoots. A

These two photos show examples of ethical photo retouching. In the photo above, the sire has perfect feet placement and an excellent head angle. Can you identify the ethical photo editing differences in the photo below?

Ethical photo alterations include: Removing telephone poles, transplanting fence line background, cleaning manure off animal, removing flies from animal, and cropping and centering the photo. No alterations were made to the physical conformation of the animal.

Unethical photo retouching involves any alteration of the animal's conformation and should not be practiced. The left photo shows an original photo. The right photo has been unethically edited. The calf owner altered the animals topline, flank, head, and hindquarter. This practice is unacceptable and gives a misrepresentation of the animal.

good photo editor or graphic designer can remove dirt, mud, manure or any unwanted objects from backgrounds and animal hides. Certainly, this type of editing is often faster and easier than trying to relocate a subject or wash and dry animals.

Anything that alters the physical conformation of an animal is completely unethical and should never be practiced. Straightening animals' top lines, cleaning-up underlines or front ends, or making adjustments to make an animal appear longer, taller, deeper bodied, etc. is never acceptable. These practices give the consumer an image that does not adequately reflect the subject's characteristics and is unfair to prospective clients.

Photo editing standards are very important for all livestock merchandisers. The goal of any producer should be to portray the animal as realistically as possible, without any alteration or artificial adjusting. Remember, the consumer has a right to see exactly what they are purchasing. Set photography standards and adhere to them; not just for your own benefit, but for that of the entire industry.

Accuracy in Show ring Accomplishments

When listing show ring accomplishments of animals, breeders should accurately represent the accomplishments of the animal. The following table is a good reference for how to correctly list show awards:

Supreme Champion or Grand Champion Overall - The one animal selected as the champion amongst all other breeds and animals in the show.

Champion – Used to designate the one winner within breeds, such as Champion Charolais Heifer. The reserve title would be Reserve Champion Charolais Heifer.

Division Champion – Used to designate the one winner of respective age divisions, such as 2-Year-Old Bulls.

Class Winner – Use to designate the one winner within respective age classes such as January 2011 bull calves. Do not use class champion, as championship names are reserved for higher awards.

Many breeds award special designations to shows such as a National Show. In this case, winners of this show are called National Champions. This title is given to the one grand champion of the show, and the reserve champion would be classified as the Reserve National Champion. Other award titles for winners at these shows would include National Division Champion, National Class Winner, National Champion Get-of-Sire, etc.

The phrase "many time champion" is one that commonly appears in livestock advertising. Just because an animal was on a show string does not make it a champion. To classify an animal as a many time champion, that animal should have been the grand champion at two or more livestock shows. Similarly, an animal that never won a championship but won several classes could be captioned as a "Consistent Class Winner" or "Many Time Class Winner."

Accuracy in Advertising Statements

Any good journalist understands the importance of fact checking and accuracy in written material. The same principle holds true in advertising. Use caution when making statements in advertising like "The first and only..." or or other definitive statements unless they can be accurately checked and proven true.

Accuracy in Performance Information

The basis of managing any kind of successful livestock operation includes the recording and reporting of factual information. The breeding of livestock, regardless of breed or species, demands a good record keeping system. Performance information such as birth weights, weaning weights, EPDs, carcass information, and genetic testing results should always be accurately represented. Many publications now require breeders to include the date that EPDs were retrieved in advertisements to preserve accuracy.

In the club calf business, particularly the promotional sire segment, there are cases where animals may be traded many times between various owners before reaching their final owner. In cases like this, information may not be transferred accurately between the various parties involved. Thus, it is better to accurately represent what information is known, instead of concocting fictitious information such as incorrect birth weights or incorrect birth dates.

The most successful livestock producers spend serious amounts of time in studying pedigrees, records, sales, and performance information. This information is only useful when it is correct, honest, and accurately presented.

Copyright and Trademark Infringement

Livestock breeders should be extremely cautious to avoid copyright and trademark infringement, especially when choosing names of animals that are identical or similar to other business trade names. Before naming any animal after a protected brand or trademarked name, the breeder should consult with the owner of that trademark to obtain written permission to use the brand. For example, the owners of the crossbred bull Monopoly obtained written permission from Hasbro, Inc. before naming the bull which granted them permission to use the name and similar design style of a Monopoly board. However, the bull owners had to list a trademark disclaimer on all printed material stating that Hasbro, Inc. owned the registered trademark of the word Monopoly.

Additionally, livestock producers should pay careful attention to avoid copyright infringements on any graphic designs such as logos, print advertisements, or banners. Never attempt to imitate another designer or breeders work without obtaining their written permission first. While it may seem like extra red tape, these precautions are extremely necessary in order to avoid legal issues or copyright infringement cases.

Ranch House Designs Livestock Advertising Code of Ethics

At RHD, we have an unwavering commitment to high ethical standards in all our business endeavors, and thus we created this livestock advertising code of ethics. We have adopted this set of livestock advertising ethics standards and welcome others to use this same code of ethics in their own livestock advertising programs.

1. RHD will maintain the highest level of truth and ethics in all advertising, public relations, marketing, news, editorial writing, and websites created.

2. RHD exercises only ethical livestock photo retouching, and will never alter the physical conformation of any animal represented in a photograph.

3. RHD exercises extra care when creating advertising materials that may be viewed by children or young adults, including using discretion based on the product, animal name, and any implications related to alcohol, prescription drugs, or guns.

4. RHD respects and maintains consumers' personal privacy regarding their preferences to receive selected marketing, news, email blasts, and advertising.

5. RHD accurately represents the circulation of printed and electronic media.

6. In the event of a potential ethical concern, RHD team members will privately discuss any ethical concerns with clients during the creation of advertising, designs, websites, or content in all efforts to resolve ethical dilemmas according to the highest ethical standards.

7. RHD strives for accuracy of all materials published, whether it be online, through social media, or in printed materials. Personal opinions in editorial writing will be labeled as such, and any errors of significance will be corrected when called to attention.

8. RHD follows federal, state, and local advertising laws and cooperates with industry self-regulatory programs for advertising practices, including but not limited to the Livestock Publications Council, U.S.D.A., U.S. Patent and Trademark office, Federal Trade Commission, Food & Drug Administration, and National Advertising Review Council.

9. RHD respects the intellectual property of all photographers and designers and abides by all copyright laws.

Conclusion

Aiding in the quest for ethical advertising practices, most publications reserve the right to reject any advertising copy that is deemed questionable or that may jeopardize the publication's mission. When in doubt, it is best to err on the side of caution, especially to avoid any case for legal concerns.

The science of selling has been described as "where reason meets desire." Advertising plays a major part in creating this desire. And as livestock producers, we must do so in an honest, true, and ethical manner.

CHAPTER 14:

Customer Service

Since the purebred livestock business is largely influenced by the people that do business together, customer service and integrity are two factors that play a very important role in establishing and retaining loyal customers.

There are many ways that a purebred livestock producer or seller can help create a good customer service protocol for his operation. It's easy — use The Golden Rule! Treat your customers as you would want to be treated. By expressing a sincere, genuine interest in your customer's success, you add even more value to your operation and to your product.

While traditional businesses have 9 to 5 hours, the purebred livestock business is basically a 24/7, 365 days-a-year business – including holidays. Sellers of purebred livestock may be called upon at any given time (night or day) by current and prospective clients who want more information or have questions about the product. Many sales inquiry calls occur early morning or late night, since these are typical hours that those involved in the agriculture sector do business. Always be prepared by knowing what you have for sale, important upcoming events of your farm, and prices of your sale animals.

Before the Sale

Many people ask, "How do you find new buyers?" Promotion is a huge tool in building name recognition and attracting new buyers, but ultimately it is the producer's responsibility to make the sale and seal the deal. The best promotion in the world is useless unless the seller can successfully negotiate, complete, and provide service after the sale occurs.

The easiest method to gain new clients and offer customer service is to be a friendly, easily-accessible person who is welcoming to everyone. People are definitely more likely to do business with someone they know, feel comfortable dealing with, and like. Sure, you may have the highest-quality livestock on the planet, but if you are hard to deal with, rude, or treat your customers badly, you won't get very far in the purebred business, and you certainly won't establish a repeat customer base.

Get involved in associations and organizations with fellow members who share your interests. Attend breed functions and shows, volunteer and serve on committees, and be positive at the events. In time this will help you build a reputation among your peers and will help you make contacts within the industry. One complaint often heard is that the "big breeders" always seem to be in the limelight. That's not by accident, but by design. The outgoing breeder works hard to get his business name out as much as possible and is highly involved in the industry.

Every operation should develop a mailing list, which should be continually evaluated and updated with former clients and those who have expressed interest in your product. How do you begin to make a mailing list? If you have a website, one should definitely have an option to 'join our mailing list' on the website. This will give you the opportunity to capture visitors' contact information. Offer a ranch guest book at your office and ask every visitor to sign in and provide an address, and, more importantly, a phone number and email address. If you host a sale, ask buyers to register for a buyer number to help get an accurate list of those who attend the sale.

Some associations also offer use of their mailing lists to members, as long as the information is relative to that group. Most times the associations will not actually give out the mailing list to protect their members' privacy; however, they will provide the mail service for a small fee. Advanced associations, like the American Angus Association, allow you to sort their mailing list into different categories such as location, area of interest, and more. Take advantage of these great tools for creating a highly targeted mailing list that reaches people who are most likely interested in your product.

Once your mailing list is established, use it! If you are hosting a production sale, send catalogs and a short personal note to the key buyers in last year's sale. Or, if you are consigning animals to a special sale, look through your mailing list and select target clients to invite to the sale. Your clients will greatly appreciate a personal copy of your sale catalog along with a short, handwritten note inviting them to the sale. This gesture only takes a few minutes per client and makes a big impression. After you send letters, an actual follow-up phone call will also show the potential customer that they are important.

When making personal connections with prospective clients, show sincerity. Everyone can spot a "used car salesman" a mile away — and most people avoid this person. Focus on building friendships and relationships... not high-pressure sales. The most reputable livestock producers often say that their livestock sell themselves and that they only provide the informational tools to help clients make a decision.

During the Sale

One often begins the sales process by making an initial contact, then providing the prospective clients with the information about the animals for sale. When negotiations begin, show the clients the animal and allow them the opportunity for close visual inspection. Show the client a copy of the registration paper and performance data and discuss the strengths of the animal. Answer any questions about the animal and provide input when asked. Show clients its sire and dam or siblings, if possible. But ultimately, let the client make his own decisions and never try to coax or pressure someone into a sale.

In the purebred livestock business, clients naturally migrate to those breeders with quality livestock and a good reputation. Integrity is part of customer service. Be honest about your livestock, their ages, their pedigrees, and their strengths. Particularly, if you are selling livestock to enter into show competition, be sure to point out the flaws of the animals. It is far better to state any areas you would like to see the animal improved up front, rather than have an unhappy customer six months down the road. There is no perfect animal, and one can almost guarantee that later someone is going to find a flaw in your animal. If you, as a breeder or seller, have openly discussed the faults of each animal prior to selling, then the buyer will be prepared for any negative comments made about the animal by judges, jealous neighbors, or competitors.

Never run down someone else's livestock. Your competition can undercut your price and product, but the harder they try to run you down, the quicker they destroy whatever goodwill they have with other breeders. While you may be tempted to point out flaws of other animals, or share negative things about a competitor's operation, focus on presenting the strengths of your animals and your operation instead. Have confidence in your livestock and convey that to your potential clients.

In many operations, especially larger ones, there may be multiple staff members or family members who are capable of making sales for the business. Keep communication lines open and try to keep a consistent point of contact for every client. For example, if one staff member shows the client around initially, that same person should show him around every time he comes. Once negotiations begin, whoever is handling the sale should make every effort to complete the sale from start to finish, without interference from other staff members (especially once price negotiations begin). All staff members should be informed and aware of pending sales, specific animals that may be under contract, or any prices that have been extended to potential clients.

Realizing that purchasing decisions may take some time, a seller can extend an 'option' to a prospective client who may need more time to consult with family members or make up his mind. A typical option might along the lines of that the interested buyer has 48 hours to make a decision, and that during that time the seller would not sell the animal to another interested person.

After the Sale

Good customer service continues after the sale is complete. Once the deal is made, record the transaction to create a written record of the sale. Many operations use sale tickets or sales books to formalize the sale. Make sure to write any important details on the sales ticket immediately following the sale so that details will be remembered down the road. Also, it is a good idea to write the salesperson's name on the sales ticket, so that person can remain the primary point of contact for the client. If the client returns to the ranch a few years later, you can also reference the sales ticket to make sure the same salesperson shows that client around when he returns to the ranch.

Payment should be made on livestock before they leave the breeder's possession. As soon as the check clears, transfer any applicable registration papers to the new owners. This makes the new owner feel good about the purchase when they receive their new registration papers of ownership in a timely manner.

The most important rule of service after the sale is this: Always be available to your clients after the sale. When your clients call you, answer the phone or return the call as soon as possible. If your client emails you, write back promptly. If you see a client in person, make a point to always stop and speak to him.

Never underestimate the impact of a personal, handwritten letter. When sent immediately after the sale, a concise thank you note is a great token of appreciation that creates goodwill between the buyer and seller. Your thank-you note can be something as simple as "Thank you for visiting the ranch this weekend. We appreciate your purchase and wish you the best success with your new show heifer from V8 Ranch." Include a copy of your business card as a handy reference for the buyer. Email, though not as personal as a handwritten letter, is a great way to communicate with clients on short notice. Personal emails are valued much greater than mass emails or email blasts.

Regularly follow up and check in with your clients. Though the old saying of "no news is good news" usually applies to selling livestock, it's still a good idea to occasionally check in with your clients just as a follow-up to show your interest in their success.

Make personal visits. A personal visit to your peers' farms or ranches is more valuable than any printed or web-based advertisement you can purchase. Personal visits help build relationships, express interest, and help you learn more about your clients' operations, needs, and goals. When traveling, look through your client list and consider stopping in for a visit at one of your client's farms in the area. Always use courtesy in scheduling farm visits, just as you would like your customers to do for you. Call ahead and try to avoid holiday visits.

Address problems and concerns quickly and efficiently. My father, who has been working with some clients on our ranch for over 30 years, has a saying he tells every single buyer at our ranch: "If you have a great experience with your calf from us, tell your friend. If you have a problem, tell me, and I'm going to make it right. If you have a problem, but don't tell me, there's no way I can correct it." This simple philosophy has been the basis of our ranch's huge client base of repeat business. We stand behind our product, and we value customer satisfaction as our top priority.

In the event that a problem arises with something you have sold, address the problem or concern immediately, and make every attempt to fix it. Though it might be easy to hit the "ignore" button on your cell phone when a disgruntled customer calls, this is the worst possible thing you can do. Listen to the customer's concern, and then let him know how you are going to correct it. Or better yet, ask him how he would like you to correct it.

If you are a seller of purebred livestock, it is highly recommended that you have some type of written guarantee in place that addresses areas such as breeding and fertility guarantees. This is a key area where the written sales ticket or written partnership agreement is useful, so that both parties can refer to the original transaction if there are areas that need to be clarified. Many breed associations also have standard guarantees or standard sales terms in place that members may use.

Customer Service Incentives

There are hundreds of ways to make your customer feel valued and appreciated after the sale; however, the majority of customer-service incentives fall into one of these five broad categories:

- Social incentives
- Financial incentives
- Management incentives
- Marketing incentives
- Educational incentives

Social Incentives

As stated throughout the book, the purebred livestock business is a people business. There is no doubt about it. While good livestock will always sell themselves, being an overall socially likeable person will most definitely help your marketing program.

Customer service means being readily available to interact with your clients. For those producers in the purebred livestock segment, it usually means nonstop service, including holidays. Those involved in marketing purebred livestock should make every effort to reach out to their clients by:

- Being willing to host prospective clients at your ranch on evenings, weekends and holidays. Since this is when many working professionals have time off, this is usually the best time for them to visit your ranch.

Also, as livestock clientele reach out to more non-traditional or urban clientele, many clients may want to come visit your ranch during the holidays as a "get away" from their hectic daily lives in town.

- Always making a point of finding and speaking with your clients when you are both at the same events, sales, or shows.

Field Days

Field days are another great way to provide social interaction for your clients – and prospective clients. If nothing else, a field day will force you to clean up your place and take care of all the odds and ends that need to be done around the farm. Carlos Guerra holds several field days at his ranch throughout the year. Their field days include activities for both adults and youngsters, ranging from educational topics and junior show incentives to showmanship contests and more. When hosting a field day, never let anyone leave empty handed. Guerra, for example, makes sure every junior who attends always leaves with a door prize. Adults should leave with a brochure, flyer, or some type of promotional material from your ranch. Guerra feels that field days help get people involved at a local and regional level, and then those friends return home and spread good words about your operation or association. Similarly, he recommends attending as many field days as possible.

Industry Organizations

One can make great contacts with current clients and meet prospective clients by attending industry events, joining industry organizations and getting involved, and introducing themselves to new breeders at events.

Mix & Mingle

It's a natural tendency to migrate towards the same old friends you've known for twenty years when you get to shows, events, and sales. Honestly, lots of times you probably drive to the sale together, sit together at the sale, and then ride home together and stop and eat along the way! While these long-lasting relationships are fabulous, it is also important to try to meet someone new at every event. If you're one of those folks I just described, consider making a pact with your old buddies that once you get to the show, you will watch at least half of the show while sitting with someone new. Or, if you just can't bear the thought of not sitting with your neighbor throughout the whole sale...at least invite someone new to sit at your table. You'll make a new friend!

Be There!

Support your clients' events and activities. If you sell a show heifer, attend a few shows to watch your client show. If this client holds a field day, be there. If your client is hosting a sale, tell others about the sale and make sure you are in attendance. Better yet, if your budget allows, bid on and possibly purchase from your clients to support their sales.

Social Networking (Online and In Person)

Accept and request friend updates on Facebook. Here's another area where children and spouses can play a part in a business customer service. When you make a new contact in the business, look them up on Facebook and request to be their friend. If a client requests to be your friend, accept them.

Make your stalls the hot spot at shows. If you're going to a show, bring plenty of chairs, drinks, and snacks. Consider setting up a small seating area for a booth at livestock shows to give your friends and customers an area to sit down and visit. Tasteful hospitality will naturally draw folks into your area and help stimulate conversation and new friendships. Everyone likes to be part of a social group of friends, and this is no different at livestock shows. Use the old adage "the more the merrier" and make every effort to welcome others into your group at livestock shows.

In short, if people enjoy being in the livestock business, and especially interacting with you, they are more likely to stay in the business. People will forget what you did and what you said, but they will always remember how you treat them.

Financial Incentives

Financial incentives are a strategy that offers extra financial benefits to customers using your product. These may include:

- Premium-based incentives for show winnings
- Financing incentives
- Accepting calf scramble certificates
- Customer loyalty or volume discounts

Premium based incentives

Premium-based incentives for show winners are very commonly used by those operations which focus on a show or youth livestock market. Topline Farm, an Angus farm in Illinois, implemented a junior rewards program in 2010 to help capture interest for their junior show heifer market.

Topline Farms offers premium incentives to their junior customers in the form of cash payment or credit to be used towards a future purchase. Their award tiers include top payments of $3000 to a junior who exhibits the grand champion female at the National Junior Angus Show or their state fair. Premiums are also available at different awards levels for breed championships, division championships, and class winners at national, regional, state and county shows.

Express Ranches of Yukon, Oklahoma, offer the most comprehensive junior incentive or scholarship program in the cattle business, called the Express Ranches Progressive Junior Scholarship program. Since its establishment in the 1990's, this ranch has provided a scholarship assistance to hundreds of their junior customers who are furthering their education. This program is also based on different award levels and also awards premium incentives to repeat junior customers.

Winners in the Express Ranches Progressive Junior Scholarship earn a 5 to 10% appreciation award that varies annually based on the current economic condition. This appreciation award earns interest from the beginning of the month in which the original award was given until the funds are properly distributed to the accredited two or four year college, university or vocational technical school.

Special futurities or shows are another incentive program that can be utilized by producers. One such example of this would be the Heritage Classic, which is a two-ring show hosted by Heritage Cattle Company each summer. In the first ring, only cattle purchased from the ranch are eligible to compete. This gives the ranch's customers an opportunity to not only get additional practice before the summer state and national shows, but also to earn prizes and premiums while competing against only those cattle raised by Heritage. The second ring of the Heritage Classic is the owned show. This show is open to those animals either sired by or out of Heritage animals. It is designed to encourage junior exhibitors to breed their former show heifers and produce their own show animals. It also serves as a tool to help junior exhibitors develop their herds.

Futurities are commonly seen through consignment sales. For example, the North American International Heifer Calf Futurity is one such show hosted by the American Shorthorn Association. This event, which is based around the Louisville Shorthorn sale, holds a special "sale show" in which the cattle consigned to the sale compete for spots in the sale order. This show is judged by a panel of ten Shorthorn breeders with a mix of both older breeders, new enthusiasts, and new breeders. The judges place the class through secret ballots, so there is no pressure or possibility for embarrassment of a new breeder who may not be confident in his judging ability. After the show, the cattle are tied in the show ring for a pre-sale viewing. This creates a social and educational opportunity for prospective sale buyers to view the cattle and also gives an opportunity for the consignors to compete in a show. It also serves as a way to excite new breeders who are able to get to serve as futurity judges.

Financing incentives

Financing options provide another financial incentive to buyers. In Brazil, their livestock auction system is based on a payout program. For example, when the winning bidder places a bid for $1000, it actually means that his monthly payment will be $1,000 for the next 24 months. So this bid, while it appears as $1,000, actually equals $24,000 when the full payout occurs. This is the expected way of buying and selling in Brazil.

In the United States, we almost always sell for one flat price; however, some producers offer the option for customers to pay for the animal through installment plans. These types of installment plans are most commonly seen in private treaty purchases, where the buyer and seller can negotiate payment terms as part of the sales transaction. If you are a prospective buyer of livestock at a public sale, the standard payment terms are always "payment due at the time of sale." Check with the sale staff and ranch personnel to make any special arrangements for purchases made in auction sales.

Another option, such as the Cyclone Trace Financing Plan, allows clients to apply for financing from the farm to be used towards the purchase of an animal in the sale.

Farm-lending agencies also offer financing options and loans for youngsters to purchase cattle through 4-H and FFA programs. Often times however, these agencies require additional paperwork, photo documentation, and copies of registration papers in order to approve the loan. Ranches that are comfortable working with these agencies have an advantage over those ranches that refuse to provide that type of paperwork or administrative support.

Calf Scramble Certificates

Many major livestock shows offer a valuable program called the calf scramble. In this program 4-H and FFA members have the opportunity to earn certificates that are exchanged for credit to be used towards the purchase of an animal. The exhibitor selects his animal from the breeder, cashes in the certificate to the breeder, and the major stock show reimburses the breeder for the certificate.

In these cases extra effort is required on both the part of the buyer and seller to adhere to program guidelines, complete appropriate paperwork, and provide required project reports. Also the seller must realize that payment from the sponsoring show will be made at a later date, usually sixty to ninety days after all of the paperwork is completed. Ranches not prepared to accept these terms should most likely not sell calf-scramble cattle; however, those who are willing to devote the extra time and effort to participating in this program can usually realize the benefits of an expanded junior clientele.

Repeat Buyer / Volume Discounts

Discounts for repeat buyers or volume purchases are popular incentives when large numbers of livestock are being bought and sold. These discounts are used by sellers to encourage volume purchases and repeat business. Buyers like these incentives, too, since they reward the buyer for loyalty.

Express Ranches, a purebred seed stock cattle operation in Oklahoma, offers several incentives in their annual bull sale. If an individual purchases five to nine bulls in the sale, they offer a five percent discount off the total purchase price. If a buyer purchases ten or more bulls, they enjoy a ten percent discount off the total purchase price. For a cattleman who buys eleven bulls at $3,500 each, he essentially buys ten bulls and gets the eleventh bull for free!

Sandpoint Cattle, another seed stock operation in Nebraska, offers discounts based on the number of consecutive years a buyer has purchased cattle from the ranch.

Management Incentives

Management incentives can provide extensive customer service and are usually a basic requirement for those doing business in the purebred livestock sector. Depending on the client's needs, management incentives include things like:

- Delivery or trucking assistance
- Halter breaking show cattle
- Providing clipping, fitting, and feeding assistance
- Providing showing assistance
- Providing hoof trimming
- Allowing clients to travel with you to shows
- Stalling clients' cattle with ranch's cattle at shows
- Assisting with hotel and airline reservations for clients
- Providing semen to breed the animal
- Breeding or AI assistance
- Training in showmanship

These small activities can add up to offer a huge incentive to buyers, especially those who are inexperienced or non-traditional livestock buyers (like urban 4-H or FFA members). These incentives are also very appealing to buyers who may not live on a farm or ranch but enjoy owning purebred livestock and appreciate the extra support that ranches may offer. This part of customer relations can get expensive and sometimes gets abused by the customer. If your ranch offers these types of incentives, again, put any limitations in writing at the time of the sale. For example, a ranch may state that they will exhibit a partnership animal at no cost at the major shows such as Houston, Denver, and Louisville but that any additional show expenses would be split between owners.

Marketing Assistance

If your customers are successful in breeding, showing, and ultimately selling their product, they will be more likely to come back to you for repeat business. Carlos Guerra has used this philosophy to build extensive marketing options for his clients. By helping them be profitable in the cattle business, he has also helped create a steady market for his product.

One such example of this is the LMC $ellabration sale, which is an auction that Guerra hosts exclusively for his clients. To be eligible for the sale, animals must be out of cattle purchased from LMC. This provides an option for their clients to participate in a nationally known sale, without having the expense or effort of trying to host an individual event. All of the LMC ranch staff helps out with the event, promotes the sale, advertises the sale, and provides a great marketing option for their clients.

The Great Shorthorn Revival offers a similar option for marketing assistance for their clients. This is a nationally-known, purebred Shorthorn consignment sale hosted by Little Cedar Cattle Company. To be eligible to consign cattle, one must have been a buyer in one of the previous Revival sales or be selling genetics from a purchase in a past Revival.

Other ranches may offer marketing programs or buy-back programs to allow their clients to commingle their calves with the parent ranch-raised calves for marketing purposes.

Assistance in advertising and promotion can also be a big incentive to buyers, especially in partnerships. For example, when selling an interest in a herd sire, buyers find it appealing that the seller will promote and advertise the sire as well as handle semen sales. Or, perhaps a ranch may sell a prominent show heifer but offer to continue to feed, care for, and exhibit the female throughout the conclusion of her show career at no cost to the buyer.

Educational Assistance

Educational assistance is a huge selling point when marketing to those who are new in the business and looking for guidance through a mentor or close friend. Often times the seller of purebred livestock automatically becomes the mentor to these new breeders as part of their customer service repertoire.

For example, junior exhibitors look to their livestock provider for information on feeding, management, fitting, care, showmanship, breeding, and more. Many times, a junior exhibitor is very cautious in making any decision without first consulting their trusted breeder. In some cases these customers may call daily with questions and may be asking for advice. They may call frequently before shows to inquire about feeding, watering, weigh-backs, what certain judges like, or other questions that may arise any time, day or night. Always answer and always do your best to help the customer. For this reasons lots of producers who market to junior exhibitors host educational clinics free of charge for their clients.

Historical information is also an extremely important educational tool that can be shared between buyers and sellers. At our ranch, my father has a historical magazine collection of different livestock publications dating back to the early 1900's. Hundreds of our customers ask him if they can browse through his collection when they visit the ranch. They also love to contact him to learn about historical bloodlines, older breeders, and other information that can only be learned by asking those who experienced it. Sharing this information with your more inexperienced clients can help them learn more about the business and develop a bigger passion for the industry as they learn its history.

Breeders may also choose to provide educational services to their customers by providing complimentary magazine subscriptions to informational publications or even paying their client's membership fee to join their respective breed associations.

Education isn't free, and this is especially true in the livestock business. Your customers trust you for your expertise and knowledge – be willing to share it.

A Word on Working with Partnerships

Partnerships in the purebred livestock sector require a heightened level of customer service to ensure that both parties maintain open communication about both the management of the individual animal as well as financial management in the partnership.

Joint ownership of purebred livestock is very common, especially when dealing with donor females, herd sires, promotional sires, show animals, and groups of cattle. Seed stock producers may also elect to sell certain genetic material such as flushes, semen packages, breeding interests, foreign breeding interests, and DNA cell lines, while still maintaining 100% of the possession of the animal.

Because of the complexity of some partnerships, we recommend that any time a sale transaction occurs, the producer writes down a formal sales ticket or sales agreement. Ideally, partnership agreements should be drafted by an attorney, but this rarely occurs in the purebred livestock business. This is understandable since many sales transactions occur at shows, on the farm, or in the barns. At minimum, both parties should write down the details of the sales transaction, including names and addresses of the partners. It is also important to jot down the major details of the partnership such as possession rights, breeding rights, show exhibition terms, etc.

Another important point to discuss is the responsibilities of each party up front, as well as how financial responsibilities will be handled. Specifically discuss who pays for care and management of the animal, promotion, advertising, and additional fees such as semen collection, flush expenses, and more. Discuss how the partners will handle decision- making issues related to management and breeding, what type of decisions can be made by the managing partner, and what requires consultation of all parties. Also discuss any specific arrangements such as retaining show rights, retaining flushes, retaining DNA cell lines, etc.

When forming any partnership, it is important that trust be established and maintained by both parties. A partnership can be either wonderful or a disaster. Make every effort to know your partner and only do business with those you trust and have confidence in.

Proper management of partnerships will help build lasting relationships and create successful experiences for both the buyer and the seller...meaning repeat business and financial success for everyone involved.

Success in the purebred livestock business depends on satisfied customers. Do everything in your power to achieve that.

Sources:
Carlos Guerra, personal interview.
Bob Hudgins, personal interview.
Jim Williams, personal interview.
Cari Rincker, personal interview.

Legal Considerations for Partnerships, by Cari Rincker, attorney at Rincker Law, PLLC. www.rinckerlaw.com

"Service after the Sale" by Patrick Wall, American Shorthorn Association. Published in The Shorthorn Country, November 2011.

CHAPTER 15:

Putting It All Together

Success in any agricultural business' advertising and public relations campaign includes the following four steps:

1. Planning
2. Implementing
3. Maintaining
4. Evaluating

"Nothing is more dangerous than an idea when it's the only one you have," is one of my and Julie French's favorite quotes from the book *A Whack On The Side of the Head – How You Can Be More Creative*, by Roger von Oech. This is a very treasured book of mine that was given to me by my father.

Regardless of where you are in the process of marketing, there is always room for improvement, and always opportunities for trying something new. Always remember that marketing, advertising, and promotion are an ongoing process that requires constant evaluation, attention and time.

Through my experiences at Ranch House Designs, I have come to realize that many agricultural producers do not have the time or capability to create their own advertising and promotional programs; many prefer to work with someone they trust, to aid them in this process. Regardless of how marketing materials are being created, either by owner, by various publications, by sale managers, or by working with a graphic designer or agency, the following checklist can be helpful in developing a marketing plan.

Checklist for Building a Livestock Advertising Campaign

Basic Foundation
- Develop a name for the business
- Develop a slogan for the business
- Choose signature colors
- Register a domain name, or have a design firm register it
- Order business cards
- Join applicable professional or breed associations

The Next Steps
- Develop a ranch logo
- Take photos of product
- Build a company website or hire a professional designer to do so
- Create a company Facebook® page
- Purchase links on industry websites
- Participate in a consignment sale or host a private treaty sale
- Send an email blast directing potential clients to company website
- Place ads in printed publications
- Print matching business envelopes, letterhead and sales tickets
- Put up a company sign at headquarters
- Order company caps or T-shirts
- Support industry activities through charitable sponsorships

Complete and Total Presence
- Display or exhibit at industry shows or conventions
- Shoot video of products and publish online
- Develop and maintain a company blog
- Develop a sire/donor catalog or sale catalog
- Publish a quarterly newsletter
- Host a production sale
- Purchase radio advertising
- Purchase television advertising

Sale Planning Checklist
(Also useful for special events / field days)

6 Months to 1 year prior to event
- Determine the sale date as far in advance as possible. Check with breed association, auctioneer, Internet broadcaster, hotel and photography / video professionals for possible dates. Don't make your event too close to another sale in your area unless you share a weekend that can be good for both parties.
- Once the date is set, list the date on your website and all ads.
- Establish a budget and schedule for advertising, photos, and catalog.
- Assign specific responsibilities and tasks and keep up with progress.
- Book necessary staff members such as auctioneers, ringmen, sale crew. Book the best sale staff available as they play a major role in sale success.
- Make any adjustments to last year's sale schedule for improvements.
- Begin plans for caterers, food, hospitality, and restroom facilities.
- Begin managing your livestock. Make sure the animals are in good shape, especially ones that will be photographed to use in advertising. Consider breeding dates and sire selection. For example, heavy bred females might be easier to handle and look better than a new mother with a young baby. A.I. service to a popular bull will outsell something bred to a lesser-known bull.

4 Months Before Event
- Ask associations or publications to add your event date to their calendar.
- Write a short news release announcing your event.
- Rent equipment including tents, sale ring, bleachers, PA system, chairs, tables, lights, rest rooms, cooking equipment, etc.

3 Months Before Event
- Complete food arrangements. Decide whether you are going to have it catered or have some of your consignors or friends help with food. Check with your local civic clubs to see if there is an opportunity to get them involved with your event. Make specific note to serve food that is desirable to event attendees, i.e. a beef dinner for a cattlemen's event.

- Begin working on your mailing list. If you plan on self-mailing your catalog, begin updating your list and adding possible new contacts. Postage and printing costs can add up, so consider the cost of self-mailing vs. inserting your catalog into a magazine such as your breed publication.

2 Months Before Event

- Begin heavy promotion of event through advertising and email blasts.
- Make personal invitations to any special guests (mail, email, phone call)
- Check with your vet on what health tests will be needed and set him up to do what is needed with plenty of time.
- Follow up with any rental companies to confirm dates and arrangements.
- Order staff apparel such as staff shirts, jackets or caps.
- Finalize printed materials (catalog, program of events, signage).
- Video the livestock. Make sure they are in good shape and clipped.

1 Month Before Event

- Begin any broadcast advertising such as radio announcements or television advertising to promote event.
- Begin preparing facilities for event (clean up, painting, etc.).
- Continue to make personal contact with your repeat and potential buyers by letter, phone, email blasts or personal visit.

Week Of The Event

- Send email blast reminder of event.
- Determine a plan for name tags for the event if desired.
- Make extensive personal phone calls to invite buyers. By this time, one should have a good idea as to who is going to buy your animals. If you have done your homework, you will know who these people are.
- Finalize facility preparation. Grass mowed, pens clean, signage displayed. Significant animals displayed in visible pastures or pens. Bleachers, tables and chairs ready, restroom facilities in place. This is your first chance to make a good impression.
- Meet with event staff to assign tasks and answer any questions. Be clear on what each persons responsibilities include so no job goes unfinished.
- Make any updated printed materials such as supplement sheets, revised

programs, maps, etc.
- Complete event decoration.
- Review key animal details with sale staff and employees. Provide as much info as you can on their EPD's, performance, show record and production of parents. Make sure all staff members are informed and able to answer questions.

On the Day Before The Event
- Go through one last trial run or overview of the event with event staff.
- Display any highway directional signs to the event.
- Do a final clean up of facilities and animals. If you are displaying animals, make sure they are clean, freshly washed, and in pens with proper feed and water.
- Present the animals in the best possible manner to give a great first impression when buyers arrive.
- Hang signs identifying animals or lot numbers for sales.

Day Of Event
- Maintain cleanliness of ranch, pens, and alleyways.
- Have sale supplements sheets and pen maps available
- Register buyers / visitors (make sure to get buyer's address and association membership numbers if applicable to complete necessary transfer purposes).
- Have plenty of beverages available, especially if it is a hot day.
- Have a good experienced team in the back and in the ring helping with the cattle to keep them as calm as possible.
- Settle sale. Make sure that you get each buyer's association number for transfer purposes, address, email address and delivery instructions.
- Give buyers a copy of their invoice, pedigrees on the cattle and their health papers.
- Offer buyers a gift of appreciation (caps, working sticks, etc)

Event Follow Up
- Settle sale promptly. Complete transfer of animals.
- Help coordinate delivery and follow up with thank you letters and a call.
- Make notes of any areas for improvement for next year.

Event Staff Needs for Sales or Field Days

Prior to Event:
- Consider hiring extra help if needed to assist with facility / ranch clean up prior to, during, and after the event.

Sale Hosts:
- All ranch staff members should be available and mingling with the crowd to answer questions from potential buyers.

Registration Table:
- 2 or more individuals to greet and register buyers, provide settlement sheet, and pen maps.

On the Block:
- Auctioneer
- Knowledgeable individual to discuss individual animals
- Sale ticket clerk
- Internet bidding staff

Ringside:
- Ringmen
- Sale ticket runner (to run tickets from clerk to settlement table)
- Internet bidding staff
- Phone bidding staff

Settlement Table:
- Settlement staff (2 or more individuals to make for speedy settlement)
- Livestock insurance representative
- Veterinarian or person designated to handle health certificates

Sale Crew / In The Back:
- Staff to keep pens clean and animals fed during event
- Staff to make sure animals are clean before entering the sale ring

- Staff to work sale ring entry gate / exit gate
- 1 person in the ring to help move livestock around
- Staff to keep livestock moving in the back and sale running efficiently.
- Load Out: Knowledgeable people who are detail-oriented who can work efficiently to keep both cattle and customers calm during load out process. It helps to have a separate crew for load out who are fresh and ready to work after the sale, since others who have worked the sale are tired.

Hospitality:
- Hospitality team to make sure water coolers are stocked, food is available, rest rooms are clean.
- Individuals assigned for meals.
- Individuals assigned for snacks and beverage service - especially replenishing drinks. It is certain that a sale will always run out of bottled water in the heat of sale weekends.
- Person assigned to check rest room facilities periodically to make sure they are clean, in working order, and stocked with supplies.
- Person(s) to manage parking, either through posting signs or directing spectators to parking spaces if necessary.
- If parking area is a far walking distance from the event, it is recommended to offer transportation such as on golf carts or other small vehicles to accommodate those who may not be comfortable walking long distances.
- Staff responsible for decorations.
- Is your ranch kid friendly? Make sure you have drinks suitable for children like bottled water or juice packs. A kids play area with simple toys like balls, toy trucks and digging equipment can occupy kids for hours. If you really want to knock it out of the park consider a swingset or inflatable bounce house to keep the kids busy while the parents look at the livestock.

Special thanks to Carlos Guerra, Julie French, Martha Garrett, and Merridee Wells for their contributions to this checklist.

The Buyer's Checklist

Adapted and updated from The Simmental Shield, July 1983.

Before making a livestock purchase, here are some important items to consider:

- Visually analyze the animal for structural soundness.
- Study the performance record of the particular animal.
- How does the animals sire and dam look? Perform?
- Research the sire and dam's performance data (EPDs, sire summary)
- Ask about any genetic tests or markers available on the animal
- If this animal has had offspring, look at their performance and sales records. Check them for structural soundness. Are they the kind you want?
- Are any full siblings or half siblings available? Are they the kind of livestock that you want? Are they uniform? What do their progeny look like and how do they perform?
- Is there carcass information available on the animal, progeny, or parents?
- What is the animal's disposition like?
- Does this animal's genetics fit your breeding program and goals?
- Has the animal had all its vaccinations and meet your herd health program?
- What is his condition?
- Is he healthy? Has your veterinarian checked the animal out?
- What is the reputation of the breeder who owns the bull?
- Does the sellers breeding and management program relate to yours?
- How has this animal been raised? (I.e. in a pasture, in the feedlot) Does this fit with your plans for him?
- Is the animal registerable, and has he been registered? Have you seen the registration papers? Will the seller transfer ownership to you?
- Is the animal a guaranteed breeder?
- If you have to cross state lines, do you have all of the documents you will need?
- Based on what you've seen, are you paying enough for the animal, or paying too much? In other words, are you buying an animal that's "good enough" or one that will take your program to another level?

Livestock Style Guide

RHD follows AP (associated press) style for advertising and editorial content, but in many cases, AP style does not include livestock terminology. This style guide was adjusted from the Livestock Publications Council stylebook and updated with terminology commonly used at RHD. Some of the original entries were created from various sources, and when these entries are listed they are followed by an abbreviation of the organization that made the rule: Kansas Livestock Association (KLA), The Quarter Horse Journal & The Quarter Racing Journal (QH), Paint Horse Journal (PHJ), Hereford World (HW), National Swine Registry (NSR), Texas A & M University (TAMU), and the Wyoming Business Council (WBC).

adjusted weaning weight – In beef cattle, a calf's weight at weaning adjusted to a standard 205 days. (TAMU)

adjusted yearling weight – In beef cattle, a calf's weight at 1 year of age, adjusted to a standard 365 days. (TAMU)

age – follow AP style – use figures for all ages: (cattle and people) Example: The 3-year-old bull is by Mytty in Focus. Elmo was a champion at 2. Now 8, Elmo is enjoying his retirement in Texas.

agribusiness – One word. Production on the farm and ranch. Storage, processing and distribution of agricultural commodities. The manufacture and distribution of farm supplies. (KLA)

AI sire – A bull who's semen is available for us in artificial insemination. Prefer the term AI sire rather than promotional bull or promotional sire.

Ak-Sar-Ben – A large (but not considered a major) livestock show held in Omaha in the fall. Ak-Sar-Ben is Nebraska spelled backwards. The A, S and B are capitalized.

American breed– Term used to describe a beef animal that has Brahman influence or has the appearance of Brahman influence and developed in the United States.

American Royal – A major livestock show held in Kansas City, Missouri each fall. Should be listed as American Royal in advertising captions however is commonly referred to as "The Royal" or "Kansas City" amongst livestock people.

Animal Names/Prefixes/ Private Herd Numbers –

• All herd prefixes are capitalized. For example JDH Madison de Manso

737/4

- Many breeders use Mr. or Miss in front of their names. Take careful note of whether a period is used or not after Mr. Some breeders include a period, others do not.
- Look at each individual animal to determine if there are spaces between the prefix letters or not. Some Angus animal names have a space between the prefix letters, while most other breeds do not.
- Cap ET when referencing embryo transfer calves. In international animal pedigree, TE is used to represent embryo transfer. Example #JADL El Rey TE 110. TE represents that he is an embryo transfer calf, his name is not El Rey Te.
- Animal names should not appear in all caps, however when you retrieve a listing of pedigrees from a breed association data file they will be in all caps and will have to be changed. If used in a sale catalog, all caps are acceptable, but not preferred.
- If using a nickname, spell out the full name of the animal then follow the nickname in quotes. Ex: Mr. V8 287/5 "Superstroke". The nickname can be used on all future references.
- If referring to an animal who primarily goes by his private herd number, spell out his full name on the first reference, then go by his private herd number on future references, as long as there is no other animal with that same private herd number in the same story.

appendix – the former term used to described Shorthorn influenced cattle in their Appendix Registry. Now referred to as ShorthornPlus

artificial insemination – AI stands alone. No periods. OK to use AIing, AIed. Never use A.I.ed. Example: The gilt was bred AI. (NSR)

average daily gain (ADG) – ADG stands alone on first reference. The daily change in body weight gained by an animal per day. (HW, KLA,NSR)

backfat – One word. Layer of fat between the skin and muscle along an animal's back (measured in inches) (HW, NSR)

bald face – Two words. White face, including eyes and nostrils. (PHJ)

baldie, black baldie, red baldie – Lowercase. A color, not an official breed. (WBC)

Beef Improvement Federation (BIF) – A national beef cattle organization dedicated to performance testing and genetics. Can be abbreviated BIF on second reference.

birth weight, weaning weight, yearling weight – Two words. (NSR). Acceptable to abbreviate as BW, WW, YW in all advertising.

bloodline – One word. (PHJ,QH)

bottom side – referring to the dam's side of the pedigree

bred-and-owned – Refers to the owner's relationship to the cow and calf. Use hyphens. (KLA)

breed names – Always capitalize the bred names of sheep, beef cattle and horses. "Horse" is capitalized ad dictated by breed associations: Paint Horse, Quarter Horse, Morgan Horse. Other species are not typically capitalized: Angora goats, Holstein cattle. (HW, QHJ, PHJ)

British breed – Breeds of beef cattle originating from England. (Angus, Red Angus, Shorthorn, Hereford, Polled Hereford)

brockle-face– A descriptive term for an animal who has splotches on its face. For example, a black brockle-face steer would have a black body and a white face with splotches of black on the face. Similar to a baldy but not the same.

calf – The young of cattle. Plural is calves. (TAM)

cattleman, cattlemen (pl) – One word (WBC)

cattlewoman, cattlewomen (pl) – One word (WBC)

champion titles – Do not capitalize champion, division or class titles in editorial show results. (HW). Capitalize when referring to national and international proper titles or in captions or example, Miss V8 100/7, the 2010 National Champion Female. *Note, many advertisers or clients will disagree with this, because they feel it should be capitalized. In advertising, do as the client says. In editorial, do not capitalize.

club calf– A specific niche of the show cattle business in which cattle are marketed for show purposes only, usually with a terminal endpoint in junior market shows. Many club calves are crossbred cattle. The club calf business is one sector of the beef industry, and should not be referred to as the 'club calf industry.' It is not an industry, it is a niche market of the total beef industry.

cold pedigree – An animal without a known bloodline or pedigree

conformation – An animal's physical characteristics – often misspelled "confirmation." (WBC)

cow – A sexually mature female bovine. Usually called a heifer until its first calf is born. (TAM, WBC)

cow-calf – Hyphenate in all instances. (HW, KLA)

cow herd – Always two words. (HW)

crossbred, crossbreeding – In general, the mating of different breeds in the same species. Do not capitalize. (NSR, TAM)

CWT – Hundred weight. Stands alone. No periods when abbreviated. (HW, NSR)

dam – In livestock, a female parent. (NSR, TAM)

donor / donor female – A female used in an embryo transfer program in which her embryos are flushed and then implanted in recipient females. The term donor is used to signify higher quality animals because usually donor females are elite females of which their owners want to proliferate their genetics. Sometimes you will see people refer to an animal as the 'donor dam' in which the term donor is just being used to show people that the mother is a donor cow.

Ear notch, ear notching – Always two words. (NSR)

Ear tag, ear tagging – Always two words

embryo transfer – (E.T.) The process of transferring a donor embryo into recipient cow. E.T. can stand alone. Example: She is an E.T. calf. (HW, KLA)

expected progeny difference (EPD) – EPD can stand alone in editorial. Plural is EPDs (no apostrophe). Note: positive (+) is assumed; print (-) if a negative. Example: BW -3.0, WW 27, YW 57, M 10, M&G 18 (HW, WBC)

F1 – The first generation of offspring or progeny from the parents. (TAM). No hyphen (RHD)

flushmate – One word. Animals who are full brother or full sister and conceived out of the same embryo transfer flush.

foreign – Avoid this use, instead, use international when referring to other associations, breeders outside of the United States, etc. If used in advertising some breeders will want to say the term "Semen for sale foreign only" and you should try to get them to use the term international.

full sister, full brother – Two words. An animal with the same sire and dam. (QH, WBC)

FWSSR – Fort Worth Stock Show and Rodeo, a major livestock show held in Fort Worth, Texas in Janury/February. FWSSR is a generally accepted abbreviation among livestock people, however if listed in ad captions we prefer to list it simply as "Fort Worth": i.e. Class Winner, 2012 Fort Worth. Generally the "Stock Show and Rodeo" part is left off of livestock captions.

get-of-sire – A group show division in which the animals shown are sired by the same bull. Hyphenate between words. (HW)

granddam – One word. (HW, QH)

granddaughter – One word. (HW)

green – A term used to describe an animal that has been handled or trained on a limited basis.

grey or gray Brahman – GREY is the official spelling according to ABBA registration papers and ABBA Standard of Excellence. Gray is acceptable, but officially it should be grey. Capitalize when referring to an official title such as 2013 International Champion Grey Brahman Bull but do not capitalize in any other use.

half sister, half brother – Two words. An animal that has the same mother or father, but both parents are not the same. (QH, WBC)

halterbroke – Means the animal will accept a halter and can be led. (PHJ)

hair show / haired show– A term used to describe a livestock show in which beef cattle or market steers are allowed to be fit and have no restrictions on the length of their hair, compared to shows like Houston which are slick shows and the animal cannot have more than ¼ inch of hair.

heifer – A young bovine before she has her first calf. (TAM)

herd sire or herd bull – Always two words. (HW)

highest-selling–the single animal that was sold for the highest price at a sale. (RHD) Also, can use the terms second-highest-selling female, third-highest-selling bull, etc. (QH)

HLSR – Houston Livestock Show and Rodeo, a major livestock show held in Houston, Texas in February/March. HLSR is a generally accepted abbreviation among livestock people, however if listed in ad captions we prefer to list it simply as "Houston: i.e. Class Winner, 2012 Houston or say Class Winner, 2012 Houston Livestock Show. Generally "and Rodeo" is left off of livestock captions because it is irrelevant to the livestock sector.

Hoodoo–A term used to describe a specific line of Charolais cattle originating from the Hoodoo Ranch in Wyoming.

horned, polled – Lower case when describing am attribute of an animal "He is a polled Brahman bull." (RHD) Capitalized if used in the name of a farm, ranch, sale, show or other proper usage. Examples: Curtis Polled Herefords. The Smith family raises horned Herefords. (HW). Polled should never be used in all caps in editorial, even though some breeders do this in their ads (RHD).

International Champion – Used to describe the one single grand champion of the International Brahman Show. If reporting show winners from the International Brahman Show, refer to the titles as International Calf Champion, International Class Winner, International Reserve Senior Champion.

in vitro fertilization – The reproductive process in which an egg is fertilized by semen outside of the body. Acceptable to use IVF on first reference and as a stand alone term.

judge – The official person designated to serve as the official of the show and place the livestock. Do not capitalize.

Limousin / Limi – Cattle breed. Prefer to use the entire word and not just "Limi" to describe the breed.

Livestock judging team – A group of four to five high school or college students that typically evaluate groups of hogs, cattle and sheep at livestock judging contests, and defend their class placings with oral reasons. Never hyphenate. Never capitalize. (NSR)

Maine-Anjou – Cattle breed. Always use a hyphen between the words. Prefer to use the entire word and not just "Maine" to describe the breed. Their % registry breed is called MaineTainer. Capitalize the letter M and the letter T – one word.

major / major show– A term used to describe the largest stock shows in the U.S. These shows are commonly known as Denver, Louisville, Kansas City, Fort Worth, San Antonio, and Houston. In Texas people refer to the "Texas majors" as Fort Worth, San Antonio and Houston, and sometimes Dallas (State Fair) in the fall.

many time champion - Careful of using this phrase in editorial or advertising copy without verifying that the animal has actually been a champion multiple times. This caption should only be used for animals that have been grand champion (the single champion) at 2 or more shows. If the animal has not been a "many time champion" alternative titles can be used as "member of ranch showstring" "competitive showring contender", or list their highest placing ever attained.

NAILE – North American International Livestock Exposition, a major livestock show held in Louisville, Kentucky each November. This show was created to replace "the international" after it moved from Chicago. NAILE is a generally accepted abbreviation, however if listed in ad captions we prefer to either 1) spell it out as North American or 2) list it as Louisville. For example, Class Winner, 2012 North American or 2012 Louisville Class Winner.

National Cattlemen's Beef Association (NCBA) – The organization formed out of the consolidation of the National Live Stock & Meat Board/Beef Industry Council and the National Cattlemen's Association. (HW)

National Champion – Used to describe the one single grand champion of the National Brahman Show. If referring to the United States National Champion, you do not have to specify U.S. National Champion. However if referring to the national champions of international countries, add the country name first, i.e. Honduras National Champion, Panama National Champion. If reporting show winners from the National Brahman Show, refer to the titles as National Calf Champion, National Class Winner, National Reserve Senior Champion.

National Shows – In some breeds, this is the one show designated each year as the "national show". This can be capitalized when describing the entire official event such as "The 2010 National Brahman Show will be held in Dallas." Some breeds have more than one national show, in a type of rotating format or regional format.

natural service – Allowing livestock to breed naturally. (TAM)

nicknames – Use quotes around nicknames of animals and people on first references. (QH). If using an animal's nickhame in editorial, you can include only his nickname on all future references. i.e. Mr. V8 287/5 "Superstroke" was one of Jim Williams favorite bulls. When Superstroke showed, he was very gentle. (RHD)

NWSS – National Western Stock Show, a major livestock show held in Denver, Colorado each January. NWSS is a generally accepted abbreviation, however if listed in ad captions we prefer to either 1) spell it out as National Western or 2) list it as Denver. For example, Class Winner, 2012 National Western or 2012 Denver Class Winner.

off side – The right-hand side of an animal. Compared to the "show side" which is the left hand side while looking at the animal. (TAM)

open – A non-pregnant breeding female.

outcross –Used to describe animals of different genetic strains or pedigrees but within the same breed. Usually people use the term to describe animals that are not related to the most popular bloodlines at the time, i.e. Ali is an outcross to Heat Wave. Outcrossing (verb) is the practice of introducing unrelated genetic materials into a breeding line to increase genetic diversity.

out of – An animal breeding phrase that refers to an offspring's mother or dam. Example: Calf B is out of Cow A. (TAM). Never say "Out of" a bull, use by or sired by. (RHD)

PHA (PHAF or PHAC) – PHA is a genetic disorder that is found a lot in club calves and therefore many club calf cattle are tested for this disorder and the results of their tests are listed in their advertising, simply as PHAF (meaning PHA free) or PHAC (meaning PHA carrier).

phenotype – The physical appearance and traits of an animal as determined by its genetic makeup. (NSR)

polled – An animal that is hornless through heredity. (TAM) Also see horned/polled listing.

pounds – In editorial use abbreviation in all cases (lb.). Express as singular or plural in writing as it would be spoken. Ex. He bought 10 lb. of flour. The bull weighed 1,500 lb. It was a 1,250-lb steer. Birth weight: 97 lb. (HW)

produce – A female's offspring. Note: a cow "produces," a sire, "sires." (PHJ)

progeny – Offspring of one or both parents. (NSR, TAM)

program names – Capitalize the names of special award programs. Example: Show Heifer of the Year. Ex: She was named Grand Champion of the junior show and Reserve Champion of the open show. (H, NSR)

promotional bull – A term commonly seen in the club calf business to identify a sire that will be heavily promoted as a candidate for AI. We prefer to use the term AI sire vs. promotional bull because promotional bull can be seen as too much of a hyped up term, and have negative meaning.

purebred – Describes an animal that has been bred true to its breed for many generations and has a registered, traceable pedigree. (NSR).

recip / recipient female - The surrogate mother of a calf through an embryo transfer program. The official spelling is recipient female, however you commonly see recip used throughout the industry.

Red Brahman – Capitalize when referring to the official breed, for example, 2012 International Champion Red Brahman Bull. Do not capitalize when using as a color descriptor, i.e. Winchester Magnum 999 is a red Brahman bull.

regions – Southern California, West Texas, East Texas, South Texas. Refer to AP stylebook for many cases, remembering to exercise some restraint. Consult coworkers when in doubt. (QH)

registered – Do not capitalize registered, except when it is part of a farm or ranch name. Example: He has 48 registered Shorthorns. WHR Shorthorns is a family operation.

registration number – Generally isn't used in editorial copy, except for foundation horses and King P-234. P-171,000 is the last registration number with a "P" prefix. (QH)

rib eye area or ribeye area – Varies among publications. In pork, refers to the carcass area at the 12th rib. Ribeye measurements are used to determine yield and quality grades. (HW, KLA)

seedstock – Registered animals used to start a breeding herd. (TAM)

ShorthornPlus – The herd registry of the American Shorthorn Association for animals that are Shorthorn crossed cattle. Used to be called their Appendix Registry.

showbox – A piece of equipment in which exhibitors store their tack. One word. Also the name of The Showbox, a youth livestock magazine based out of Texas.

show divisions – do not capitalize age divisions (calf, junior, senior) or experience level divisions (amateur, open or youth) unless used as part of a specific title: She is the 1993 Youth High-Point Breakaway Roping Champion. (QH) In the senior division, Miss V8 341/7 was named champion. (RHD)

show cattle, show bull, show heifer, show string, show ring, etc. (QH) – Two words

show steer – Two words. A show steer is a market animal designed to compete in junior market shows with a terminal endpoint. See club calf.

SimAngus – A hybrid beef cattle breed made up of Simmental and Angus. Capitalize the S and the A. One word.

Simmi – A slang term used to describe the Simmental breed of cattle. Always spell out the word and do not use this abbreviation. If the abbreviation is used, it should be used with two M's to reflect the original spelling of the breed, not one m.

slick shear – A term used to describe the act of removing the hair from a beef animal, usually for the purpose of show competition or heat tolerance. In Texas, many shows are "slick shows" meaning the cattle are not permitted to have more than ¼ inch of hair. Houston and San Antonio are slick shows compared to Fort Worth, which is a haired show.

smoky – A color description used to refer to cattle that are of the cream or greyish color, or a mixture of both. Usually these are cattle that have some type of Charolais influence. The official spelling is smoky, however you will see it incorrectly spelled as smokie. It can also be described as smoky-colored.

state fairs – Each state's annual competitive agricultural event with rich histories dating more than 100 years in most cases. Capitalize when referring to the proper noun such as Illinois State Fair. Do not capitalize when describing the general event such as "We went to the state fair on Tuesday. Vs. We went to the Iowa State Fair on Tuesday. (RHD). Note that many state fairs are called simple the ____ State Fair such as Iowa State Fair,

Illinois State Fair, etc. However some are referred to as the State Fair of
______ as in the State Fair of Texas. Also important to note that in Texas,
some of their regional fairs are referred to as 'state fairs' such as the West
Texas State Fair (Abilene) however the one official state fair is in Dallas.

steer – A male bovine castrated as a calf. (TAM)

Texas majors - A term used to describe the largest stock shows in Texas
people, specifically Fort Worth, San Antonio and Houston, and sometimes
Dallas (State Fair) in the fall.

TH (THF or THC) – TH is a genetic disorder that is found a lot in club
calves and therefore many club calf cattle are tested for this disorder and
the results of their tests are listed in their advertising, simply as THF
(meaning TH free) or THC (meaning TH carrier). It is acceptable to just
list these abbreviations.

weaning weight– Two words. (NHF, NSR). Acceptable to abbreviate as
WW, YW in all advertising (RHD).

yearling weight– Two words. (NHF, NSR). Acceptable to abbreviate as YW
in all advertising (RHD).

yield grade – Lowercase. Always capitalize with the numeral (1,2,3,4 and
5). Yield grade is based on a calculation that estimates the percentage of
boneless retail cuts from a carcass. (HW)

AFTERWORD

Advice for Aspiring Livestock Marketing Professionals

Through the years, many young women contact me with advice on how to build a career like mine in the livestock business. It is a compliment and an honor, and I am always happy to share my experiences with others. For this second edition, I added this afterword to share some of the advice, and things I have learned along the way, with others who may be interested in a career like mine.

First and foremost, I have been blessed with a unique creative ability that is only given by the true Creator, GOD, for which I am truly thankful. My grandparents, Sloan and Mollie Williams, own V8 Ranch, and my parents, Jim and Luann Williams, have worked there since 1976. I moved back home to the ranch in 2006, and today my husband and I also both work at the ranch as well. As a child, I recall watching my family negotiate cattle deals with purebred Brahman breeders around the world. Ultimately, through the years I have performed virtually every ranch marketing duty from barn work, showing cattle, learning and talking about bloodlines, creating the ranch web site, handling the ranch advertising, to eventually negotiating those big-dollar cattle sales I had watched my grandparents and parents make while I was a youngster.

That foundation, combined with education and experience, gives me the ability to help my fellow livestock producers create the absolute best advertising and promotional programs for their ranches. Quite simply, I have built one of the best livestock advertising agencies in the world because I know the business from the ground up. I learned by watching others who are successful in the business and do it with integrity.

Livestock advertising and promotion is a fun and challenging endeavor, yet the fundamentals of livestock advertising will always be rooted in a firm principle that does not change: focus on good livestock. The good ones make my job as a livestock marketing professional easy and enjoyable.

I also think that RHD has been a success because we don't just talk the talk. We know what it's like to get up at 2:00 a.m. to get cattle ready for a show. We've been the ones leading cattle to tie outs in the freezing rain. So myself as well as the people I hire on my team understand our clients in all aspects....from marketing to sales to even the challenges that ranchers face. For example, I understand if it takes my cattle clients about a month to pay me for their bull sale catalog design because they are having to wait to settle the sale. I get that. Or if they are needing an urgent banner for Denver because they changed out what calves they are taking to the show. It's happened to me and I understand. I also have learned to develop a thick skin when working with clients because my husband and my dad are pretty typical of every client I work with -- they know their business and they expect me to know mine. They don't have time for a lot of big talk, and they want results. And that's my job, to deliver.

The people you surround yourself make a big difference.

There are certain people that helped put me where I am: my husband Brandon; my parents, Jim and Luann Williams; grandparents, Sloan and Mollie Williams and Gene and Sammie Johnston; special friends, Gary & Kathy Buchholz; and my Agriculture Science teacher, Dean Fuchs. In 1991, Kathy let me help her design the booth for Buchholz Bros. Shorthorns at Louisville, and I fondly remember that she let me write the animal names on the booth in bubble letters. It was an old-school felt board where we velcroed the pictures on the felt. I was 11 years old at the time. Who knew that would be my first display design. Even more surprising, for anyone who knows Mrs. Buchholz, is the fact that she would allow bubble letters on their display. Later, attending Texas A&M University, I had the honor of working with Dr. Larry Boleman and Dr. John McNeill, two great Extension educators who greatly impacted my life.

People talk about the Aggie network, and it don't get me wrong - it is powerful, but the Michigan State network is just as strong in the ag community, and I am convinced that my 2 years at MSU just took me to a whole new level professionally. There I had the opportunity to learn from Dr. Dave Hawkins and Dr. Harlan Ritchie, as well as interact with so many other top young people who were students at MSU at that time. It was truly the mecca for young people in the beef industry at that time and I am thankful that this Texan decided to move up to Lansing for those 2 years.

I have a fond appreciation for the confidence that my parents gave me as a young person growing up in the cattle business, and I know that without that confidence I wouldn't be where I am today. When I was 10 years old, my dad would send me in the show ring with a 2,000 pound Brahman heifer who was good enough to win the Houston open show. (This is the equivalent of Denver / Louisville for Brahman people). And we would win. When I was a teenager, Dad would have cattlemen from around the world asking him questions about our cattle, and if he got overwhelmed with people he would just say, 'Ask Rachel, she knows." What kind of Dad has that much confidence in a little girl to give her that opportunity? My dad. And I can't thank him enough for that.

Never once has anyone in our family ever told my sister or I that there was something that we couldn't do because we were a girl. We were always treated like we could be anything, and do anything in the cattle business because we were smart enough and talented enough. And yes, that meant we were also able to load up feed and do cow work just like a boy. Without this confidence, I wouldn't have had the courage to start a woman-owned business in a predominantly male business. Nor would I have had the confidence to tell cattlemen what they should do in their advertising....and have them listen.

Without a doubt, the team at RHD is what makes us a success. Yes, I started it, but the team here is what makes it work. I am grateful to the team members, past and present, who have contributed to the overall development of our agency through the years, specifically key staff members through the years like Carole Arriaga, Jamie Bloomberg, Nena Boettcher, Ashlei Cooper, Ashley Culpepper, Joelynn Rathmann, Emily Forgason, Stacey Forgason, Christa Guidry, Jessica Hobbs, KC Kinder, Jackie Lackey, Lacey Lively, Jeanie Long, Catherine Neumayr, Luke Neumayr, Tricia Potts, Stacey Shanks, Tana Townsend, Elizabeth Wilcox, Kacey VanDeaver, Kimber Welch, Halle Yancey, and many more. Each of these people brought unique talents at different stages of the agency's progress, and helped establish us as the leader in livestock advertising.

Organization is crucial.

People also ask how do we keep track of everything here at RHD? On any given day we have between 100 and 200 projects on the books. The answer is our organization. We pre-schedule projects and deadlines 6 weeks out in advance so we know what projects are due when. We have Monday mornings to catch up from the weekend then we have staff meetings on Monday afternoon and get lined out for the week. Usually we start the week out with about 20 projects due that week. But by Wednesday we try to have all our proofs out so that we can have all our deadlines on track by Friday. We are very structured. That doesn't work for some potential clients who always need last minute things -- but in general, our organizational system is very efficient and keeps us all on track. We have never missed a deadline in 15 years.

We maintain one spreadsheet, called our "deadline sheet" which is our best friend, and our enemy. This spreadsheet is sorted by project deadline and contains every project we have on our books, who is the project manager, and who is the designer. Anytime any member of our team works on a project they log it there, so that all team members have access to know the status of any project. We also have project folders for each project. These are envelopes that contain all of the related materials for the project. Our project folders are color coded based on the type of project (ad, website, sign).

When I first started RHD, I was on my own. I would work 20 hours days if I needed. I carried my laptop with me every single place I went in the case that someone would email me and need their website updated that I could complete the update immediately. When I became a working mom, I made a rule that RHD was for the hours of 8:00 a.m. to 5:00 p.m. on Monday through Friday and everything else was my family time. And I have been very fortunate to be able to stick to this rule nearly every single day of the last 3 years. I learned to become more efficient at my time. I don't visit a lot in the break room and I usually grab a fast 10 minute lunch at my office. But a career in livestock advertising is a great one for a young woman, because it offers flexibility and options that make it easy to be both a working mom and a business professional.

You are your own brand.

For many aspiring designers, you start out as a freelance designer, or work somewhere but may have the ultimate aspiration of one day being your own boss or having a firm like Ranch House Designs. One thing I wish I would have known 10 years ago is that when you are a young professional - especially a young woman - you are your own brand. People will make assessments of you based on the way you dress, the way you act, who you work with, who you go to dinner with, where you socialize, and everything else. I will be the first to say that as a young professional, I made many mistakes that I wish I could go back and do-over. The key was learning from the mistakes and not making them again..that is the hard part! As my good friend Chris Boleman says, "After you have made a mistake, learn from it, and then move on to the next play. The future is brighter than the past." I also had to realize that sometimes, it is necessary to step away from a popular crowd in order to maintain your professionalism. This was a difficult thing I had to learn, but in the long run, a very wise decision for my professional reputation. It's hard enough for young women to build a name for themselves in a male-dominated industry, and it is downright near impossible for grown cattlemen to have respect for a young woman who is more worried about what concert she is going to that night or how big her bling earrings are instead of being focused on the business responsibilities at hand.

Understand the value of your time.

One student recently asked me, what do you wish you would have known before starting your career? I wish I had known the value of my time. Lots of times young designers think that they only bill for the actual time they spend designing the ad -- which may be 2-3 hours. But there is a lot more that goes into it, for example the time to organize the materials, download the photos, come up with a headline, find a background photo, send proofs, make corrections, send the file to the printer, etc. By the time it's all said and done you might spend 10 hours on an ad. At $150 per ad then you just valued your time at $15/hour. You could make that at McDonalds. So that is something you have to establish up front, and don't sell yourself short.

Should you do work for free? When you are young, this might be tempting. I did a lot of work for free but it was for organizations and student clubs in which I was a member - like Block and Bridle. The dilemma a lot of designers face is that you need design samples, but you might not be at the level where someone is willing to pay you for your services. My recommendation is to do any pro-bono work that you can for non-profit organizations or clubs. I also did free work for my own family's farm, because I could use them as my guinea-pig. This makes a win-win situation for both parties. Doing any other jobs for free is a slippery slope. Obviously your long term goal is to build a network of paying clients, thus I advise against doing free work for other businesses, because you are just setting yourself up for being expected to do it for free forever. Also, set reasonable boundaries for what projects you can handle and turnaround time you offer.

Cultivate mentor or sponsor relationships

Make connections with others in the business, preferably through a mentor or sponsor. I have been blessed to have many: Cherie Carrabba, Martha Garrett, Julie French, and others. At some point in my career, I worked for each of these talented women and the opportunity to learn from these masterminds was priceless.

When I started out in this business, I admired Christy Collins. I remember back in the late 1990s we had a Shorthorn bull named Bonfire and at one point our partner Don Cagwin suggested we get Christy to take his picture. I about hit the floor I was so excited that SHE was going to take OUR bull's picture. Through the years, we have built a solid mutual respect for each other and become good friends. But at no point during my career have I ever thought, I want to steal Christy Collins clients, nor did I ever think "I want to be Christy Collins". I wanted to be Rachel Williams. I wanted to build my own reputation based on the creativity and brain power I could offer...not trying to imitate someone else.

I find that many times today, when someone new gets in the business the first thing they try to do is 1)steal someone else's clients or 2)bad mouth the top people in the business or 3)undercut the pricing of the top people in the business. I would advise against all three of these practices. My advice is to make the best possible friends you can with the top people in the business. Ask them questions and listen to what they have to say. Learn from their mistakes so you don't have to make the mistakes yourself. Chances are established professionals in the business would appreciate a new colleague, they would be happy to mentor a young designer, and they might even be able to send some business your way.

A mentor or a sponsor won't develop overnight, and usually a mentor will naturally connect to someone they feel has a lot of potential. Last year, I started working with Cally Thomas, and I instantly thought that she was most talented young ladies I have ever worked with in the cattle business. I wouldn't say I'm her mentor, but she was so talented that I wanted to work with her and be around her. It was only natural that we would begin to work together and share ideas. To me these type of relationships are beneficial to both parties, because I can also learn from her. If you are looking for a mentor, try to exude professional qualities that would make someone want to mentor you.

It takes more than someone who can just design a cool looking ad.

A common misconception is that to be successful in the business you have to have this really cool looking ad. That's actually the least important skill in my opinion. Understanding of marketing principles, theory, sales, and understanding visual quality and pedigrees of animals are all more important for a career in this business. Good writing is also a skill that is very valuable in our business. If someone is a good writer, they can be trained to do basically any aspect of the business. But if someone can't write, then it is an uphill battle from the beginning.

There are thousands of graphic designers worldwide who could design a cool ad....probably thousands who can design one better than me. The advantage that people like me have is that we pull all the skills together to look at the big picture. It's not just about design. So when you have the opportunity to choose college courses to take, take as many diverse communications and marketing classes as possible. Take writing classes, public relations campaigns classes, video production, anything and everything you possibly can.

In closing, I want to express my sincere appreciation to the thousands of clients who trust me and the RHD team to handle their advertising and marketing. It is a job we take very seriously, because we know your livelihood depends on it. People ask me, why would you write this book, or hold your livestock marketing workshops to give away all your secrets? There is no secret to what we do. We are all in this industry together, and it is my sincere hope that this book will be a tool to help my fellow livestock producers, fellow designers, and aspiring designers find ways to improve their own advertising programs and ultimately boost ranch sales and exposure.

God Bless! -Rachel Cutrer

Topic Index